A Whaling Cruise to Baffin's Bay and the Gulf of Boothia, and an Account of the Rescue of the Crew of the *Polaris*

Albert Hastings Markham

Cambridge
University Press

CAMBRIDGE
UNIVERSITY PRESS

University Printing House, Cambridge, CB2 8BS, United Kingdom

Published in the United States of America by Cambridge University Press, New York

Cambridge University Press is part of the University of Cambridge.
It furthers the University's mission by disseminating knowledge in the pursuit of education, learning and research at the highest international levels of excellence.

www.cambridge.org
Information on this title: www.cambridge.org/9781108072007

This edition first published 1874
This digitally printed version 2014

ISBN 978-1-108-07200-7 Paperback

The original edition of this book contains a number of colour plates, which have been reproduced in black and white. Colour versions of these images can be found online at www.cambridge.org/9781108072007

CAMBRIDGE LIBRARY COLLECTION

Books of enduring scholarly value

Polar Exploration

This series includes accounts, by eye-witnesses and contemporaries, of early expeditions to the Arctic and the Antarctic. Huge resources were invested in such endeavours, particularly the search for the North-West Passage, which, if successful, promised enormous strategic and commercial rewards. Cartographers and scientists travelled with many of the expeditions, and their work made important contributions to earth sciences, climatology, botany and zoology. They also brought back anthropological information about the indigenous peoples of the Arctic region and the southern fringes of the American continent. The series further includes dramatic and poignant accounts of the harsh realities of working in extreme conditions and utter isolation in bygone centuries.

A Whaling Cruise to Baffin's Bay and the Gulf of Boothia

First published in 1874, this illustrated work by Albert Hastings Markham (1841–1918) recounts his experiences aboard the Dundee steam whaler *Arctic*. Markham also gives an account of the rescue of the crew of the American vessel *Polaris*, crushed by ice in 1872 during its attempt to reach the North Pole. The work is enhanced by details of meetings with Inuit, encounters with polar wildlife, oceanographic observations, and meteorological events. Appropriately, fellow naval officer and explorer Sherard Osborn (1822–75) wrote the introduction: he had a long interest in Arctic exploration, advocated the benefits of using steam ships in icy waters, and encouraged Markham to embark on the whaling cruise. The appendices include a 'list of birds shot', as well as data on botanical and geological specimens. Also reissued in this series are Markham's *The Great Frozen Sea* (1878), *Northward Ho!* (1879) and *A Polar Reconnaissance* (1881).

Cambridge University Press has long been a pioneer in the reissuing of out-of-print titles from its own backlist, producing digital reprints of books that are still sought after by scholars and students but could not be reprinted economically using traditional technology. The Cambridge Library Collection extends this activity to a wider range of books which are still of importance to researchers and professionals, either for the source material they contain, or as landmarks in the history of their academic discipline.

Drawing from the world-renowned collections in the Cambridge University Library and other partner libraries, and guided by the advice of experts in each subject area, Cambridge University Press is using state-of-the-art scanning machines in its own Printing House to capture the content of each book selected for inclusion. The files are processed to give a consistently clear, crisp image, and the books finished to the high quality standard for which the Press is recognised around the world. The latest print-on-demand technology ensures that the books will remain available indefinitely, and that orders for single or multiple copies can quickly be supplied.

The Cambridge Library Collection brings back to life books of enduring scholarly value (including out-of-copyright works originally issued by other publishers) across a wide range of disciplines in the humanities and social sciences and in science and technology.

A WHALING CRUISE TO BAFFIN'S BAY AND THE GULF OF BOOTHIA.

A

WHALING CRUISE TO BAFFIN'S BAY

AND THE GULF OF BOOTHIA.

AND AN ACCOUNT OF THE RESCUE OF THE CREW OF THE "POLARIS."

BY ALBERT HASTINGS MARKHAM, F. R. G. S.

COMMANDER ROYAL NAVY.

THE "POLARIS.

WITH AN INTRODUCTION BY

REAR-ADMIRAL SHERARD OSBORN, C. B., F. R. S.

LONDON:

SAMPSON LOW, MARSTON, LOW, AND SEARLE,

CROWN BUILDINGS, FLEET STREET.

1874.

CHISWICK PRESS:—PRINTED BY WHITTINGHAM AND WILKINS,
TOOKS COURT, CHANCERY LANE.

DEDICATION.

TO THE RIGHT HONOURABLE

SIR H. BARTLE FRERE, G. C. S. I., K. C. B., D. C. L.

President of the Royal Geographical Society.

DEAR SIR BARTLE FRERE,

THE permission to dedicate this narrative of a whaling cruise to you, as President of our Society, is specially gratifying to me, because it encourages me to hope that you will treat this effort to further the great cause of Arctic Exploration with indulgence, and that you will overlook the numerous shortcomings of so inexperienced an author. My intention is to convey, to the utmost of my ability, accurate information respecting the operations of that fleet, commanded by daring and adventurous seamen, which annually navigates the Arctic seas.

It is an auspicious circumstance that, in these days, as in the days of old, a distinguished Indian Administrator should be one of the leading advocates of Arctic discovery.

Many of the names in the Arctic regions, especially those at the head of Baffin's Bay, including that of the strait which will eventually lead us to the Pole, recall memories of the founders of our Indian Empire.

Smith Sound is named after the first governor of the East India Company, Jones and Wolstenholme Sounds and Cape Dudley Digges are called after the most active promoters of Indian trade; and Lancaster Sound after that gallant seaman who made the first voyage to India for the old company.

Moreover, several of our predecessors in Arctic exploration gave their lives for the service of the East India Company.

John Davis fell fighting for it in the Sea of Japan, and that glorious old navigator Baffin was killed whilst taking sights on an island in the Persian Gulf. Indian and Arctic navigators united very heartily in the olden times, and the connection between India and the far north at the present day is, if possible, even more intimate.

For the skilful seamen with whom I was shipmate this summer, and the ryots of Bengal, are engaged in two branches of the same industry, the welfare of the one depending very closely upon that of the other.

The jute manufactory, which gives employment to many thousands of industrious ryots, is dependent for its existence on the success of the whale fishery, animal oil being necessary for the preparation of the fibre. So that it is peculiarly appropriate that a statesman who has devoted his whole life to the welfare of India should now be among the foremost in advancing the cause of Arctic discovery.

Earnestly trusting that the efforts of the Council of our Society, under your Presidency, to secure the despatch of an Arctic exploring expedition may bear fruit, if not in 1874, at least in 1875, I remain, with warmest wishes for your success, my dear Sir Bartle,

Yours very sincerely,

ALBERT H. MARKHAM.

H. M. S. "SULTAN,"
Lisbon.
December, 1873.

CONTENTS.

CHAPTER XV.—ADMIRALTY INLET.

CHAPTER XVI.—PRINCE REGENT'S INLET.

CHAPTER XVII.—FURY BEACH.

CHAPTER XVIII.—CAPE GARRY.

CHAPTER XIX.—HOMEWARD BOUND.

LIST OF ILLUSTRATIONS.

Separate Engravings on Wood.

Engravings in the Text.

MAP.

INTRODUCTION.

THE following narrative, by Commander Albert H. Markham, of a whaling voyage in the Arctic regions in the Dundee steam whaler "Arctic," will, I feel sure, be read with much interest by all who relish an unadorned tale of adventure,[1] and awaken surprise amongst the large section of the British community who take note of progress in Arctic exploration. The voyage of the "Arctic," apart from the boldness and skill with which her dashing captain, William Adams, pursued his mighty and valuable prey through ice, storm,

[1] Captain Markham was appointed to H. M. S. "Sultan" before he had been three weeks in England, after his return from the Arctic regions; and his journal has been printed, in his absence, almost exactly as it was written from day to day, when the incidents it records were fresh in his mind.

and dangers, proves in a remarkable manner how much the introduction of steam power in whaling ships has reduced the risk formerly incident to navigation in Baffin's Bay and Barrow's Straits.

We have the "Arctic" committing herself voluntarily to be beset in Davis's Straits until there were some fifty miles of heavy pack between her and open water; and then, when no more whales were to be found, gallantly fighting her way by steam power through the interlaced ice-fields until the clear sea was again reached. We find that Middle Ice, which for half a century has been the bugbear of the whale fisher, when tackled by a vessel of 500 tons and 70-horse power engines, no longer spoken of as an impenetrable barrier. The whaler under sail thought himself fortunate in traversing it once in every three years, with a vast expenditure of labour, in from a month to sixty days. The "Arctic" and her sister vessels have now for nine years consecutively got through this middle ice under steam in as many hours. We see the "Arctic," in quest of her prey, passing point after point during a summer cruise, which for fifty years had been the extremes reached by discovery expeditions. Steam power has robbed the navigation of those regions of nearly all its difficulties and much of its risk. The "Arctic," with her keen hunters of the whale, dashes boldly past John Ross's farthest in 1818; Sir Edward Parry's farthest, in Prince Regent's Inlet, in 1825; Franklin's winter quarters at Beechey Island are reached; Sir James Ross's farthest at Leopold Island in 1848 visited; and many another bay and headland in those remote regions, which have taken seamen in sailing vessels, years of toil and hardship to attain, were seen and visited by Commander Markham in a summer's holiday. It was no exceptional season in which this was done. The "Arctic," under Captain Adams, has often made the same cruise, to the profit of her owners and crew, and returns again in 1874 to repeat the voyage.

When, in 1850, during the search for Franklin, I had the good fortune to command the "Pioneer," which, with the "Intrepid," were the two first steam vessels employed in the Arctic seas, I foresaw and foretold the great revolution which steam was about to effect in diminishing the risk of Arctic navigation. It was impossible for us, shackled as we were to our clumsy sailing consorts (H. M. S. "Resolute" and "Assistance"), to do much in illustration of my belief; but the revolution has come even quicker than anticipated, and from the shores of Great Britain, as well as from those of the dominion of Canada, strongly fortified steamers dash annually into the frozen seas in search of the seal and the whale, and reap rich and profitable harvests, without any risk of life; and rarely with the loss of a steamer. Steam yachts are now following in their wake, and Spitzbergen is becoming the summer field of many of our boldest sportsmen; and a trip to the Polar regions is thought as little of to-day as a hunting excursion to Norway, or a visit to Iceland, was a few years ago.

Apart from all these facts, which a perusal of Commander Markham's narrative will bring vividly home to the reader, there will be found a synopsis of the remarkable voyage made by the United States discovery vessel "Polaris," under the late Captain Hall, up Smith's Sound, at the head of Baffin's Bay. This information he was able to procure from the officers of that ship, who were picked up and brought home from her wreck by our whalers during the past summer. It will be remembered, that when England had completed her part in solving the fate of the Franklin expedition—a search which culminated in the voyage of the "Fox," and in Sir Robert McClure's great achievement in passing from the Pacific to the Atlantic, which laid open the entire geography of the regions between those two oceans—Arctic discovery was for a while allowed to rest.

But it was not so with our brethren in the United States. They, fired by the achievements of British explorers, and anxious to secure to their countrymen the honour of being equally bold and enduring, sent forth expedition after expedition under Kane, Hayes, and, lastly, Captain Hall, in the "Polaris," with a view to penetrate the great unknown area around the Pole by way of Smith's Sound. Of these expeditions the most remarkable was the last. In 1871, Captain Hall entered Baffin's Bay in the month of August in the "Polaris," a small, weak-powered steamer, by no means well fitted for the work, with a scratch crew composed of eight Germans, nine Esquimaux, thirteen Americans, one Englishman, one Irishman, and one Scotchman. From Cape Shackleton, where the ice is usually met with, in lat. 73° 30′ N., the "Polaris" sailed and steamed without interruption to 82° 16′ N., a distance of 526 miles, and was then only stopped by loose ice. The crew of the "Polaris," when subsequently witnessing the way in which the "Arctic" steamed through similar ice, acknowledged that a properly equipped steamer could have passed through the barrier which stopped their little vessel.

Those on board the "Polaris" saw the strait extending before them, with much open water and land to the north and west, which they believed lay in latitude 84° N., or within 300 miles of the pole of our earth. Wintering near their furthest point, they found abundance of animal life, saw much drift-wood of recent date, which must have come there across the Polar Sea from the shores of Siberia; and they found a tide coming from the same direction, and report that the temperature during the winter was considerably milder than had been experienced by American and English expeditions which had had, on former occasions, to winter in more southern latitudes. After the death of Captain Hall, the men in this expedition only thought of returning home, and were making fair progress in that direction, when by an accident in the autumn of 1872, a number of the crew were swept away from the ship on an ice-field, which eventually carried them down to the shores of Labrador. They were saved by one of our sealers. The remainder of the crew ran the ship on shore at an Esquimaux settlement near the entrance of Smith's Sound, abandoned her in the spring of 1873, with the intention of retreating in their boats to the Danish settlements in Greenland, but were picked up on their way by an English whaler and so brought home.

We have therefore, from their information, certain proof of the navigability for steamers of Smith's Sound, of the facility with which a retreat can be effected without loss of life, of considerable resources existing there in animal life, of land lying close to the Pole, and, from the tide and driftwood, of a water communication across that great unknown area.

In an appendix attached to this volume will be found an elaborate report, to which nearly every scientific body in this country has given its adhesion, on the great advantages to science which an exploration of the Polar area would render; and the information brought home by Commander Markham proves how right the Royal Geographical Society has been in recommending Smith's Sound as the route on which a Polar expedition should be sent, and shows at the same time how large a measure of success may be anticipated, as well as the comparative immunity from risk of life.

Never was a fairer field open to English seamen and adventurers to reap high renown and to keep our country in the vanguard of geographical discovery; and I cannot believe that that spirit which was awakened under the great Elizabeth can have passed away in the reign of Queen Victoria; but fervently trust, if our Government shrinks from its undoubted duty, that private individuals will secure to us the glory of being the first nation to have traversed the Polar Sea, as we have been foremost in all that is great and glorious in so many other parts of the earth.

SHERARD OSBORN,
Rear Admiral.

LONDON, *December*, 1873.

CHAPTER I.

THE DUNDEE WHALERS.

THE Arctic Regions have always been associated in my mind with that mysterious and indescribable longing which is usually connected with all things difficult of attainment, and therefore most desired and sought after.

Here perilous and exciting adventures await those daring spirits who may devote their lives to the acquisition of knowledge.

Hither our brave whale-fishers have annually ventured for many years in search of that huge and valuable cetacean, the capture of which is at once so hazardous and so profitable.

In short, the Arctic Regions teem with everything that renders travel exciting, and offer charms to the lovers of adventure such as are rarely met with in any other part of the globe.

It was, therefore, with no small amount of pleasure that I hastened to avail myself of an opportunity which presented itself of proceeding in a whaler to the Arctic Regions for the purpose of gaining experience in Arctic navigation, of witnessing the methods of handling steam-vessels in the ice, and of collecting information respecting the state of the ice in the upper part of Baffin's Bay; which might prove useful should an exploring expedition be hereafter despatched from this country to the unknown regions of the North.

While engaged on this interesting service I kept a journal, noting each event as it occurred, not trusting to or putting any faith in memory, but jotting down any incident the moment I could find an opportunity.

I am now anxious to convey to those who are interested in the subject as clear an idea as I am able of the work on board a whaler, and of the perils and vicissitudes of a whale-fisher's life; and it seems to me that the best way of attaining this object will be to reproduce my journal as it at present stands, to which will be added some further particulars respecting the modern whale fishery, and an account of the rescue of the survivors of the ill-fated American exploring ship, "Polaris." Although Scoresby, half a century ago, gave very full details respecting the whale fishery in the Spitzbergen seas, there have been many changes since his days, caused by the invention of new appliances and the introduction of steam. The story of the northern fishery, as it is now practised, has not hitherto been told in any detail.

In introducing this narrative to my readers I cannot do better than quote the words of Master George Beste,[1] the biographer of "Martin Frobisher's Three Voyages for the Discoverie of finding a Passage to Cathaya by the North West," who, in his dedicatory epistle to Sir Christopher Hatton, apologetically remarks that his "intente is, more to sette out simply the true and plaine proceeding and handling of the whole matter, than to use circumstance of mere words, or fyne eloquent phrases wherein if I should once goe about to entangle myselfe, it would doe nothing else but bewray my owne ignorance, and lack of schole skyll. Therefore, of me there is nothing else to be looked for, but such playne talke and writing as souldiers and marriners doe use in theyr dayly meetings and

---

[1] Master George Beste was the lieutenant of Sir Martin Frobisher's ship.

voyages, and this of necessity must anye man use that will deale with such a matter as thys is, although he were curious to the contrairie."

Before proceeding with my narrative, it may be interesting to dwell briefly on the progress of the whale fishery, which, in Great Britain alone, has been prosecuted for more than 250 years.

Until the commencement of the present century this lucrative trade was entirely confined (I here refer solely to the capture of the *mysticetus*) to the waters around Spitzbergen, which are commonly called the Greenland fishery. At the present time, with only *one* exception,[1] the vessels engaged in the Davis' Straits and Baffin's Bay fisheries are from the port of Dundee, which place, during the last eighty years, with fluctuating success, has carried on this important branch of commercial enterprise.

During what may be considered the infancy of the whale fishery, various Acts of Parliament were passed by the Legislature for the encouragement of the trade, and further support was given by the Government, which offered a bounty of as much as thirty shillings per ton on the burden of each ship employed in the fishery. At the early part of this century, and during the time those talented and eminently successful whale fishers, the two Scoresbys,

[1] This year a Norwegian steamer was up Baffin's Bay.

were employed in the trade, there were no less than one hundred vessels fitted out and despatched from different ports in England, of which Hull, London, and Whitby were the principal, and more than half that number from ports in Scotland, of which Aberdeen, Leith, Dundee, and Peterhead may be mentioned as the chief. In those days a cargo of forty or fifty tons of oil would amply compensate for the expenses of the voyage, and ninety or one hundred tons would be regarded as a most profitable and remunerative cargo. At the present time, in consequence of the enormous expenses incidental to steam-ships (all vessels employed in this trade have now the advantages of steam-power), the price of coal, and the enhanced value of all commodities, unless a ship returns with a cargo of at least seventy or eighty tons, her captain meets with anything but a warm and hearty reception from his employers. The trade at the present time, at least as far as regards Great Britain, appears to be entirely monopolized by Scotland, no port in England contributing a single ship for the prosecution of the whale fishery, and Dundee and Peterhead being the only two ports in Scotland.

No less than ten fine and powerful steam whalers leave Dundee annually for the fishery in Davis' Straits, all commanded by experienced and intelligent men. (See next page.)

They usually proceed to sea about the begin-

LIST OF THE DUNDEE WHALING FLEET.

Ship's name.	Commanded by	Owners.	Tons.	Horse power.	Rig.	Extreme Length.	Extreme Breadth.	No. of Boats.	When built.	If converted, and when.	No. of Whaling Voyages.
						ft.	ft.				
Arctic	Wm. Adams	Messrs. Alex. Stephen and Son, Dundee	439	70	Ship.	157	30	8	1867	Built for the trade.	7
Eric	J. B. Walker	Messrs. Antony Gibbs & Sons, London	412	70	Ship.	155	30	8	1864	,,	8
Ravenscraig	Wm. Allen	Mr. Lockhart, Kirkaldy	394	60	Barque	130	30	8	1853	Sailing vessel, converted 1866.	7
Tay	Wm. Greig	Arctic Seal and Whale Fishing Co., Dundee	360	60	Ship.	132	26	8	1850	Sailing vessel, converted 1857.	15
Esquimaux	Chas. Yule	Dundee Seal and Whale Fishing Co.	436	70	Ship.	157	30	8	1865	Built for the trade.	8
Camperdown	John Gravill	Dundee Seal and Whale Fishing Co.	424	70	Ship.	150	30	8	1860	Built for the trade.	13
Narwhal	Wm. McLellan	Dundee Seal and Whale Fishing Co.	434	60	Ship.	150	30	8	1858	Built for the trade.	15
Polynia	David Kilgour	Dundee Seal and Whale Fishing Co.	358	60	Ship.	146	29	8	1861	Built for the trade.	12
Intrepid	Wm. Soutar	Tay Seal and Whale Fishing Company.	326	50	Barque	126	30	7	1852	Sailing vessel, converted 1865.	8
Victor	John Edwards	Tay Seal and Whale Fishing Company.	278	36	Barque	107	28	7	1847	Sailing vessel, converted 1863.	10

ning of May, and, after passing Cape Farewell, a fortnight or three weeks is devoted to what is called the south-west fishing, in the neighbourhood of Frobisher's Straits to the northward of the coast of Labrador. Afterwards the whalers proceed up the east side of Davis' Straits and Baffin's Bay, to Melville Bay, that dreaded and in many instances fatal locality. If successful in making a passage through this hazardous bay, the ship emerges into the north water, when her course is altered to the westward as much as the state of the ice will admit, until she arrives on her fishing-ground at the entrance of Lancaster Sound or off Pond's Inlet. Here the fishing is prosecuted with energy until July, when the whales are sought for up Lancaster Sound as far as Prince Regent Inlet. The whales, whose habits are most migratory, are then followed, during the months of August and September, as far south as Home Bay and even as far as the Gulf of Cumberland, when the approach of winter warns the captain of the danger of remaining too long in those sterile and inhospitable regions of snow and ice. The return of the whaling fleet may be looked for about the beginning of November. Occasionally ships may arrive at an earlier date, as in the case of the "Arctic," which in 1871, in consequence of unparalleled success, returned to

Dundee during the middle of August, and was this year home by the middle of September. These are, however, rare and exceptional occasions. As a general rule they seldom return until five or six months after their departure.

The ships engaged in the whale fishery are all most substantially built, doubled and fortified by the application of timber and iron both inside and out. On each side of the stem are angle irons, or pieces of iron so placed as to protect the fore-foot from injury when coming into contact with heavy floes and also enabling the ship more readily to cut or break through any ice of a penetrable nature. The whalers are generally ship-rigged, which I consider to be a great mistake,[1] a barque being far more handy, and they vary from three to four hundred and fifty tons burden. The "Arctic" is a vessel of 439 tons, having a 70-horse power engine. She is the property of that eminent shipbuilder, Mr. Stephen, of Dundee.

In consequence of the reputation which the commander of the "Arctic" had acquired for dash and enterprise, I selected her as the ship in which I was to behold the grandeur of Arctic scenery, the deep fiords of Greenland, the enormous glaciers

[1] As soon as the whalers arrive on their fishing-ground the mizen topsail is invariably unbent and stowed away.

stretching for miles into the interior, the majestic icebergs, sailing along in every fantastic shape and form, and all those numerous sights only to be met with in the frozen regions of the North.

Having arranged terms with her wealthy and prosperous owners, I found myself on the 2nd of May installed on board, having signed articles as second mate to the effect that I engaged myself "to serve on board the good ship 'Arctic' on a voyage from Dundee to Greenland or Davis' Straits, and seas adjacent, for whale and for other fishing, and back to Dundee;" and, further, that I agreed to "conduct myself in an orderly, faithful, honest, and sober manner, and to be at all times diligent in my respective duties, and to be obedient to the lawful commands of my said master." The daily allowance which I should receive of butter, cheese, oatmeal, bread, beef, pork, flour, tea, sugar, lemon-juice, water, and other stores, were previously read to myself and the whole crew at the shipping office. My wages were to be one shilling per month, and I was to receive in addition the sum of one penny for every ton of oil brought home in the ship, and one farthing for every ton of whalebone. The reason that I was appointed to such a responsible and lucrative situation was, that as the whalers have no licence to carry passengers, anyone wishing to travel in a whaler is compelled to sign articles as one of the crew. That ceremony completed, he is free to do as he will, and to enjoy himself according to his own fancy and inclination.

Our captain is a kind, jovial, good-tempered man; and these qualities, added to his daring enterprise and great success during the time he has commanded the "Arctic" at the seal and whale fisheries, render him deservedly popular with all ranks and classes, and enable him to enlist a good crew, with efficient and experienced officers.

Our ship's company are a fine sturdy set of fellows, evidently intending work if they get the opportunity. A fourth part comes from the Shetland Isles, one man is English, one a Norwegian, and the remainder are Scotchmen, principally Highlanders. One of the Shetland men is both deaf and dumb, but is most willing, and is a fair sailor. He rejoices in the appellation of "Dummy."

Ships employed in the whale fishery have a complement of men according to the number of boats with which each ship is provided, so that when actually engaged in the capture of the huge monsters of the deep, all boats may be despatched, fully manned, leaving two or three hands on board for the purpose of working the ship. Our complement in the "Arctic," including the captain and myself, was fifty-five souls.

In order that all the men on board may take an

interest in and use their utmost exertions for the furtherance of a successful issue to the voyage,

Rating.	No. in each rating.	Monthly pay.		Oil-money per ton.			Bone-money per ton.		
		£	*s.*	£	*s.*	*d.*	£	*s.*	*d.*
Master	1	8	0	1	4	0	7	0	0
				up to 100 tons.					
				2	0	0			
				after.					
Mate [1]	1	5	0	0	7	3	0	10	6.
Second mate [1]	1	3	15	0	7	0	0	10	6
Doctor	1	2	0	0	2	0	0	2	0
Speksioneer [1]	1	2	15	0	7	9	0	10	6
Engineer	1	7	0	0	7	3	0	10	6
Second engineer and blacksmith	1	3	5	0	3	6	0	10	6
Carpenter	1	3	10	0	3	6	0	6	0
Carpenter's mate	1	2	15	0	2	6	0	4	0
Harpooneers	2	2	15	0	6	9	0	10	6
Loose harpooneers	2	3	5	0	3	6	0	10	6
Cooper [1]	1	3	10	0	6	9	0	10	6
Ship-keeper	1	2	10	0	6	9	0	10	6
Boatswain [2]	1	3	0	0	2	6	0	5	0
Skeeman [2]	1	3	0	0	2	6	0	5	0
Boat-steerers	6	2	10	0	2	6	0	5	0
Line-managers	8	2	5	0	2	0	0	4	0
Cook	1	2	10	0	2	0	0	4	0
Steward	1	3	5	0	2	6	0	6	0
Cook's mate	1	1	15	0	1	6	0	2	0
Firemen	3	2	15	0	2	6	0	6	0
A. B.'s	10	2	0	0	1	9	0	2	0
Ordinary seamen	5	1	10	0	1	6	0	2	0
Boys	3	1	0	0	1	3	0	2	0
Myself	1	0	1	0	0	1	0	0	$0\frac{1}{4}$

NOTE.—The speksioneer is the officer under whose direction the whale is cut up. This word is derived from the Dutch "spek," meaning blubber. Skeeman is the officer who superintends between decks the "stowing away" of the blubber in the tanks.

[1] Also harpooneers.

[2] Also boat-steerers.

each individual, according to his position in the ship, is paid a per-centage upon the amount of cargo brought home, their regular wages being small. This will account for the ease with which a successful whaling master is able to man his ship. The statement on page 11 comprises the usual scale of pay allotted to officers and men serving on board whalers.

In addition to the above wages there is also a *bonus* paid to each one of the crew composing the boat that first strikes a whale, provided the fish is afterwards captured, and is what is termed a "payable fish."[1] The harpooneer in this case receives 10*s*. if he gets fast with the gun harpoon, 10*s*. 6*d*. if with the hand harpoon, and £1 1*s*. if with both, each member of the boat's crew receiving 2*s*. 6*d*. This is termed "sticking-money," or "fast-boat money."[2] It is also usual for the master to receive £1 for every payable fish struck by a *bonâ fide* harpooneer, and £3 for those struck by a loose harpooneer, whose position may be looked upon as a harpooneer in embryo, or on probation,

---

[1] By "payable fish" are meant all those fish whose whalebone is six feet and over in length.

[2] No "sticking-money" is awarded for the capture of a fish whose bone is under six feet. The whalebone is regarded as the most valuable part of a fish, its price at the present time being about £500 per ton! A large whale of ten or eleven feet bone will yield a ton of whalebone.

and who generally holds that office for a couple of years, or until he has proved himself, to the satisfaction of his captain, to be a willing, persevering, and dexterous officer.

The duties of the crew of a whaler are allotted according to the abilities and qualifications which the men possess as fishermen. Thus the harpooneers attend to the conning of the ship during their watch on deck; the two mates and the speksioneer (who are also harpooneers) are in charge of the watch; the boat-steerers attend to the ropes on the forecastle; the line-managers to those round the mainmast and also to the cleanliness of the ship between decks; the boatswain, who is also a boat-steerer, is held responsible for all work aloft; the skeeman, who is the head lineman and also a boat-steerer, is responsible for everything between decks, and it is also his province to superintend the stowing away of the blubber. The remainder are the fore-mast hands, who make themselves generally useful where required. So that, to sum up our crew on board the "Arctic," we have eight harpooneers, including the mates and speksioneer, eight boat-steerers including the skeeman and boatswain, and eight line-managers. The rest of the crew man the boats. When all boats are away from the ship, there remain on board the captain, doctor, engineer, ship-keeper, cook, and steward. Each boat (we were supplied with eight) has a crew of six men, five rowers and one to steer. No rudders are fitted to the boats; a steer oar, in consequence of the rapidity with which, by its means, a boat may be swept round, being invariably used. The harpooneer is in charge of the boat and pulls the bow oar. It is his duty to strike the fish. The line-manager pulls the stroke-oar, and it is his province, with the boat-steerer, to see the lines coiled away clear, and to attend them when running out, after a fish has been struck.

Having thus given a short account of the interior economy of a whaler, I will, in the following chapters, endeavour to describe the cruise of the good ship "Arctic," during her seventh trip to Baffin's Bay in search of that great and unwieldy cetacean for whose capture so many bold seamen annually penetrate the mysterious regions of the North.

During my short stay in Dundee, whilst negotiating the terms of my passage, I was most hospitably entertained by several of the leading citizens of that prosperous town, amongst the most prominent of whom I cannot refrain from mentioning the names of ex-Provost Yeaman,[1] Mr. McNaugh-

[1] On my return it was with great pleasure I heard that during my absence this gentleman had been elected by a large majority to represent the town of Dundee in Parliament.

ten, the Dean of Guild, and Mr. David Bruce, the manager and agent of most of the vessels comprising the Dundee whaling fleet. From these gentlemen I received the utmost attention, and was treated with the greatest kindness and courtesy, which rendered my stay in their town both agreeable and pleasant.

I take this opportunity of expressing my gratitude to them for the very kind and cordial reception which I met with at their hands.

Scott Cliffs---a favourite fishing ground.

Chapter II.

"SPANNING ON."

WEDNESDAY, April 30th, was the original day named for our departure from Dundee; but owing to delay in the arrival of our coals, our sailing was put off until Saturday, the 3rd of May.

We were ready, and might easily have proceeded to sea, on Friday, but the old sailors' superstition, with which our crew was fully imbued, against sailing on that day, deferred our departure until half-past six on Saturday morning, at which time, with steam up and colours flying at each masthead, the "Arctic," surrounded by a small crowd, and with her head pointed between the dock-gates, was lying, waiting only for the order to "go ahead" to start on her seventh trip to Baffin's Bay.

I had but just time to bid a hasty farewell and

jump on board before the necessary orders were issued—the engines went ahead, three cheers were given by those assembled on the quay, feebly responded to by the sober ones on board. A photographer on the pier-head took a hasty shot as we turned to go out, and we all felt that we had at last started on our cruise.

As a general rule, the departure of a whaler is marked by the total incapacity of the crew to perform any duties whatever connected with the ship, in consequence of the numerous parting glasses of which they have partaken with their friends and acquaintances, and the bumpers that have been drained to the success of the voyage.

The scene on board an outward-bound whaler on the eve of departure has been described to me as one most filthy and disgusting. I am very glad to be able to state that, at any rate for this voyage, the "Arctic's" crew are an exception to the general rule, the captain informing me that he had never put to sea before with such a "good and sober crowd."

Those who were suffering from the effects of over-indulgence had the good sense to keep below, where they were allowed to remain undisturbed until the effects of their carouse had worked off.

Steaming down the Tay, we stopped off Broughty Ferry, at which place the captain allowed himself a brief leave-taking with his wife; and having landed all visitors and friends that had accompanied the ship thus far, the boats were hoisted up and secured, and by half-past eight our pilot was discharged, and, shaping our course to the northward, we steamed along the red sandstone cliffs of Forfarshire. Passing Arbroath and Montrose the shore assumed a more rugged and sterile appearance; the wind, which was adverse, had considerably freshened, and our progress was in consequence necessarily slow. In order to avoid a wasteful expenditure of fuel, the captain very wisely determined upon anchoring off Aberdeen; but being unable to obtain the services of a pilot, and darkness having set in, we continued our course to the northward. At 3 A.M. the following morning we brought up off Peterhead, to await more favourable weather. Shortly after leaving Dundee, the ship's company was mustered on the quarter-deck by the captain, and those who were able to answer to their names were divided into three watches, under the first and second mates and speksioneer respectively. A few words were addressed to them, and, after giving three cheers for the captain, they were dismissed.

Monday, May 5th.—We are now fairly away. We left Peterhead at noon yesterday, the wind having subsided, and with a light S. E. wind and a beautiful clear moonlit night, passed through the Pentland Firth. This being accomplished, by way of supper I was initiated into the mysteries of "whelks," and having by the skipper's advice smothered them well with vinegar, mushroom-ketchup, and pepper, found them by no means unpalatable.

Another favourite edible of our worthy captain is a seaweed called "dulse,"[1] which is picked up in large quantities on the beach at Broughty Ferry. This is kept in a bucket of salt water on the after part of the quarter-deck, so as to enable those so disposed to refresh themselves at their pleasure. At noon to-day we lost sight of Cape Wrath, and are all looking forward to a quick run across the Atlantic. We are fortunate in having a fresh N.E. wind, before which the "Arctic" is very lively and also very wet. I have quite shaken down to the life on board.

Everything is rough but ready, and I am treated by all with the utmost civility and attention. My

---

[1] "*Dulse*" a corruption of the Gaelic *duilliog*, from *duille* (leaf) and *uisge* (water). The name is applied to several species of rose-spored Algæ, and especially to *Rhodomenia palmata* and *Iridæa edulis*. In Ireland it is called *dillesk*. Lindley says that the old cry, "Buy dulse and tangle," may even now be heard in the streets of Edinburgh; and dulse is consumed in considerable quantities thoughout the maritime countries of the North of Europe.

messmates in the cabin are :—The master, a fine honest, good-hearted specimen of a whaling captain; James Bannerman, the mate, a strong, active, daring, and hard-working officer; Tom Webster,[1] second mate (though entered as chief harpooneer), an old and successful harpooneer, who had made upwards of thirty voyages to Baffin's Bay; Andrew Graham, our doctor, an enthusiastic admirer of nature, and a medical student at the Edinburgh University; John, our engineer, an uncouth, rough and unpolished black diamond; Peter, the second engineer and blacksmith; and Jack, the steward. Our meal hours are breakfast at eight, dinner at twelve, and tea at five. In consequence of the limited accommodation afforded in the cabin, we are compelled to have a relay of each meal. Occasionally the skipper, doctor, and myself sit down to a supernumerary meal at about 10 P. M., consisting, now the whelks have all disappeared, of lobsters, or cheese and grog, or something equally unwholesome and indigestible, but over which the skipper cracks his jokes, spins his endless yarns, and we talk over the events of the day.

It was with no little anxiety—a feeling which has now subsided to one of interest and wonder—that I

---

[1] Everybody on board the "Arctic" was called by his Christian name, a habit which I soon adopted.

watched the dexterous manner in which some of my messmates would perform the apparently impossible feat of eating eggs with a large knife. Forks are decidedly at a discount, every one going on the principle that fingers were made before that useful article. If we happen to have a joint on the table with the smallest particle of a bone protruding, this is at once seized by the hand of the carver, whilst large junks are cut off and handed to us. On the whole, our living is rough, but certainly substantial. It is amusing to remark the different degrees of "home sickness" with which my messmates are variously affected. Some are in the highest spirits, apparently pleased at the idea of being their own masters and getting away from the thraldom of a jealous and ill-tempered wife; others are in the lowest depths of despondency, and one confidentially informed me, a few hours after our departure from Dundee, whilst talking of his better half, that the "puir bodie would tak' on so," and that by that time "she would have cried a pint of tears."

To-day is what is termed "Store day," when each of the crew has served out to him his share of provisions, excepting meat, generally to last for two or three months.

The men are not put into regular messes, as on board of a man-of-war, but each man receives his own allowance, which is kept either in his sea-chest or bunk.

Cooking is allowed to go on all day and all night, and the galley fire is never let out during the whole cruise. The relief watches are always turned up half an hour before their watch commences, which time is supposed to be devoted to the fortifying of the inner man. When a ship arrives on her fishing ground, it is not unusual for a boat or boats to be sent after whales and to remain away for eighteen or twenty hours, hence the necessity for obliging each man of the watch to refresh himself before coming on deck.

Thursday, May 8th.—An unpleasant, showery day, accompanied by a heavy head swell. The wind failing at noon, steam was raised and sail taken in. We have already passed two of the Dundee whalers, which we imagine to be the "Camperdown" and "Narwhal." It is very probable that we are now the leading ship of the fleet, notwithstanding five having sailed from Dundee before us, as two, the "Intrepid" and "Ravenscraig," have to call in at the Shetland Isles to complete their crews. This is the first day that we have been able to enjoy the services of our cook, who is only just recovering from his last day on shore. One of our crew had a slight attack of *delirium tremens* this afternoon, otherwise they are all well, and confidently look forward to the prospect of a quick and prosperous voyage. They are busily employed fitting and preparing the different tackles and purchases used for hoisting in the blubber, and also in fitting Jacob's ladders (which are small rope ladders with wooden rungs), eleven of which are always kept over the side to enable the men to jump quickly on board from the ice, or when returning in the boats.

Saturday, May 10th.—Blowing a fresh easterly gale, before which the ship is going nine and ten knots. A wet, drizzly day, with a lower temperature than we have experienced since leaving England. We may at any moment expect to fall in with icebergs, and extra men are placed to look out day and night. A collision with one at our present speed would not be very desirable. The captain does not expect to make any ice until we are off Cape Farewell. No signs of the other whalers were visible to-day. We have completely run them out of sight. The ship is decidedly not comfortable in wet weather, as her decks leak like a sieve from the continual straining; and both the main cabin and my own are in a constant state of drip, which, to say the least, is most annoying. The upper deck also at sea, even in moderate weather, is rendered unpleasant for exercise, in consequence of the ship having open water-ways running fore and aft, which appear more useful in *admitting* the water than in taking it off. These little annoyances will not, however, be of long duration, as we are making good runs, and in a short time shall have crossed the Atlantic and be in the smooth water of Davis' Straits.

Our days are monotonous enough, and I shall not be sorry when we arrive at the S. W. fishing, where, amongst whales and ice, there will be much to see and learn. At present one day is the very counterpart of another. We have the same meals to go through, generally the same description of food to eat, the same jokes by our skipper, and the same stories to listen to, enlivened occasionally by an Irish song, a recitation from Shakespeare, or a reading from Artemus Ward.

Monday, May 12th.—Last night was the most wretched and uncomfortable we have passed since being at sea. Not only was the ship knocking about in an unpleasant manner, but, not having obtained sights for three days, we were all a little anxious regarding her position, especially as we were running along at the rate of eleven knots. At 2 A.M. land was discovered on our starboard bow, a capital land-fall, and at half-past seven I was enabled to take a rough sketch (though at the distance of about thirty miles) of Cape Farewell, the Staten-huk of the Dutch, and of Cape Desolation, the Cape Farewell of the Dutch. It was a fine clear morning, enabling us to get a good view of the distant land, which appeared bold, bleak, and rugged, and seemed to consist of a number of sharp, conical-shaped black hills, covered, where the summits were not too peaked, with snow. The strong contrast of the black and white gave the land a most sublime and picturesque appearance. The wind, which had lulled considerably, was blowing directly off the land, causing the temperature to fall several degrees. We may consider ourselves very lucky in getting round the cape with such charming weather, as it is a curious fact that ships seldom pass Cape Farewell without some little touch of dirty weather. The day is really beautiful; we seem to have suddenly emerged into a totally different climate, cold, but with a bright sun and clear sky. We are now fairly in Davis' Straits, and, taking advantage of the fine weather, active preparations for the capture of whales have been in progress the whole day. All hands have been as busy as bees, employed in the operation of *spanning on*,[1] which literally means attaching the lines to the harpoons, and coiling them away in the boats. The first operation was to get the boats out, which are always stowed in board for

---

[1] From the Dutch *aanspannen*, to put the horses to the carriage.

the passage across the Atlantic; and before sunset we had six boats ready, three hoisted up on each side, to go away at a moment's notice. After the lines are served out—everybody, from the captain downwards, being employed—it is quite a race as to which boat shall be first equipped, the crew that has been most expeditious giving three cheers on the completion of its work.

The manner in which the harpoons are fitted is first with about twelve fathoms for a gun harpoon, and three for a hand harpoon, of the best

Hand Harpoon.

Gun Harpoon.

white untarred hemp rope ($2\frac{1}{4}$ in.). The end of this is spliced round a thimble in the former, and round the shank of the latter. This rope is called the *fore gore* or *fore ganger*.[1] It is stronger and more supple than ordinary rope. To the other end of the *fore gore* is spliced the remaining whale lines, of which there are five in each boat, of 120 fathoms, the united length of which is over 600 fathoms, or a little more than half a mile. These

[1] From the Dutch *voorganger*, he who goes before another.

whale lines are made of tarred rope ($2\frac{1}{2}$ in.) of the very best quality. The boats having been first thoroughly cleaned, the lines are carefully flaked down in the stern sheets, in a compartment specially set apart for that purpose, with the exception of 100 fathoms, which is flaked down in a box in the centre of the boat, called the *fore line beck*.[1] A portion of the line last put into the boat, of a few fathoms in length, is called the *stray line*. It lies fore and aft in the boat, and is always ready for running out. The *fore gore* is coiled down in a

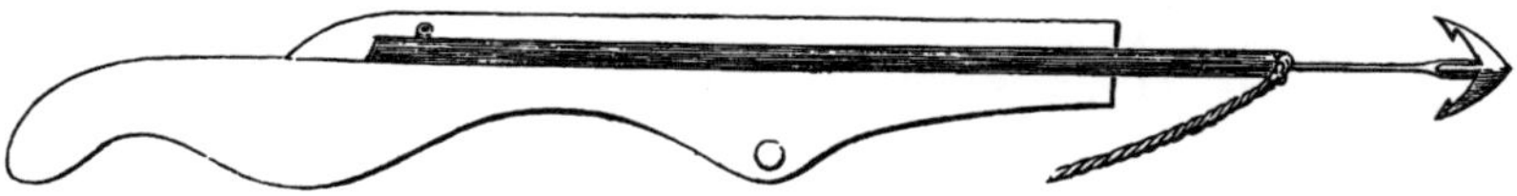

Harpoon Gun.

small tub or *kid*, which is kept in the bows of the boat alongside the gun. *Aprons* or *screens* made of painted canvas are kept over the lines to protect them from the wet, on which the boat steerer stands. The harpoon gun is fixed on a swivel in the bows of the boat, and can easily be traversed round, depressed or elevated by the harpooneer. The harpoon belonging to the gun is placed on the port side of it, and the hand harpoon on the starboard side, with its handle resting on a *mik*,[2] or

[1] *Beck*, from the Dutch *bek*, a coarse expression for mouth.

[2] *Mik* is a Dutch word for the iron in which the boom rests, crutch, &c. It is made of wood, and is used for supporting the handle of the hand harpoon.

crutch, ready for immediate use. The harpoons are made of the softest Swedish iron, which is more pliable than English, so that they may readily bend without snapping, when any strain is brought to bear on them. A bad harpoon may lose a fish whose value may exceed £1,200. Each harpoon has the name of the ship to which it belongs stamped on the shank. In addition to the articles above enumerated, the following complete the equipment of a whale boat when fully prepared for service :—

"Mik."

5 pulling and 1 steer oar.

Mast and sail (occasionally).

4 lances for killing the whale.

1 tail knife, used for cutting holes in the tail and fins of a dead whale.

1 hatchet for cutting the line, if necessary.

1 flag staff and jack, which is only displayed when a whale is struck.

1 mik or rest for the hand harpoon.

3 spare thole-pins for each thwart.

2 snow shovels.

2 piggons, or small buckets, used for baling the boat out, and for pouring water over the lines to prevent their catching fire from excessive friction.

1 marline-spike.

1 splicing fid.
1 fog-horn.
1 file.
2 boathooks.
A ball of spun yarn.
Box of ammunition, &c.

There is also in each boat a tow rope, used either for towing the dead whale or the boats employed in that duty; also a *fin tow*, which is a rope used to lash the two fins together across the belly of the fish, so as to offer no impediment whilst towing. Thole-pins are always used in whale boats, and the oars are invariably muffled by a sort of thrum mat, which lies on the gunwale. A Turk's head is worked on the loom of each oar, which prevents it from going through the grummet on the thole-pin, when pushed out after striking a fish, allowing it to run fore and aft with the boat. A steer oar is always used in preference to a rudder, over which it decidedly has many advantages. By its means, on a still day, a boat may be sculled close up to a whale, which would otherwise be frightened by the splashing of oars; with it a boat may be instantly swept round, though lying quiescent at the time; and a boat is more easily managed among ice by being propelled with its aid between floes or amongst loose pack. On striking the whale, the steer oar is pushed out to the Turk's head, which, as on the pulling oars, is worked on the loom, the boat steerer immediately throwing back the apron and kneeling down at the very extremity of the boat, where he attends carefully to the running out of the line. When a boat is in want of more line, which very frequently happens when fishing in deep water, as a whale on being struck generally dives, the boat's *piggon* is elevated on an oar or boathook as a signal—if very urgent, two, three, and even four oars, according to the immediate want, are held up.

The greatest care and attention are required from the boat's crew, after getting fast to a whale, to see that the lines run out clear, boathooks or staves being placed across the gunwale to prevent their fouling anything during the rapid career of the fish, which will often take the whole length of lines out of one boat in five minutes.

Should the line, through carelessness or from some unforeseen accident, become foul, unless immediately cut (in which case the fish is lost) the boat will be drawn down, and the crew placed in imminent danger of drowning.

Tuesday, May 13th.—Last night, as the captain, doctor, and myself were preparing our midnight meal—a villainous compound of cheese, pepper, and mustard, called a "crab"—the door opened, and M'Slasher, who had previously informed us that the appearance of the sky, which was of a greyish hue, *denounced* the proximity of ice, reported that we were passing ice.

Leaving our supper, we jumped up, and were hastily putting on our coats, when a heavy crash and a tumbling of the ship confirmed his report. We had struck a mass of straggling ice, and as we hastened on deck, large pieces were rising to the surface on each quarter; and we appeared, as was really the case, to be surrounded by innumerable masses of floating ice. Extra hands were placed on the look-out, another man to the helm, whilst another was stationed midway to pass the word along from the captain, who was directing the motions of the ship from the forecastle. During the remainder of the night there seemed to be a ceaseless repetition of the orders "hard-a-starboard," "hard-a-port," or "steady as you go;" but despite the utmost vigilance on the part of those in charge of the ship, the weather being very thick and misty, we occasionally came into collision with a mass, which would set everything on the table in motion, causing the bell to toll with a doleful and melancholy sound. This ice has probably been blown off from Cape Farewell and the west coast of Greenland during the strong easterly winds which we have lately experienced.

The morning broke fine and clear, and by seven o'clock we had steamed clear of the ice and made sail to a light northerly wind.

As the men went aloft for the purpose of making sail, we on deck were assailed by a perfect shower of pieces of ice, which having frozen on to the rigging and sails during the night, were broken off and detached from their hold in consequence of the vibration imparted to the rigging by the motions of the men.

At nine, more ice was observed ahead, and we were soon in another straggling stream; but, as it was daylight, we were enabled, though not without keeping a sharp look-out, to avoid coming into contact with the heavier pieces. It was a glorious and novel sight to me, seeing these floating masses of ice, some of them assuming the most picturesque and fantastic forms, many being of a bright blue colour. Soon the quick eyes of our captain discovered a huge seal lying on the top of one of these ice islands, dreamily looking at us, in wondering surprise, as we approached. Poor brute! a bullet from my rifle (the captain having deputed me to shed the first blood of the cruise) terminated its existence; a boat was lowered, and the monster brought on board. Others were now observed, and four boats were despatched to effect their capture, in one of which I went. Directly a seal was shot, we would at once pull in to the ice on which it was lying, and I was surprised at the marvellous rapidity and dexterity with which our men would skin, or, as it is termed, "flinch" the beast. I had the curiosity to time a couple of men whilst performing this operation on a large seal. It was

Seals on the Ice.

actually "flinched," and the skin thrown into the boat, in fifty-eight seconds! We were away in the boats about a couple of hours, during which time we obtained fourteen seals. After the skins are taken on board, the next operation performed on them is that of "krenging,"[1] which is stripping

[1] From the Dutch *kreng*, meaning carrion.

or cutting off any small portions of flesh that may be adhering to the blubber, which latter is then cut off from the skin, and this last process is called "making off."[1] The blubber is put into barrels or tanks between decks, and the skins are salted and stowed in a cask with brine.

In the afternoon the crow's nest was got up, and the remaining two boats hoisted out and equipped.

We are now fully prepared and anxious to wage war with the huge monsters of the deep.

The crow's nest is simply a large cask or barrel, which is triced up to the main-royal mast-head. The lower end rests on an iron jack above the eyes of the top-gallant rigging, secured to the mast with an iron band, and the upper part has an iron strap which goes round the royal pole. On the top is an iron framework for resting a telescope. It is altogether a very ingenious contrivance. There is a small trap-hatch in the bottom sufficiently large to admit a man, which is then shut down, and serves for the inmate to stand upon. When on the fishing ground the crow's nest is always occupied.

---

[1] "Making off," according to Scoresby, is derived from the Dutch word *afmaaken*, signifying to finish or complete; probably in consequence of its being the concluding operation. The Dutch *afmaaken* really means to kill, to give the *coup de grâce*.

In the evening we passed large quantities of ice, and saw several bottle-nosed whales; but the capture of one of these does not offer sufficient inducement to delay our progress towards the south-western fishing-ground, and therefore the captain resolutely holds on.

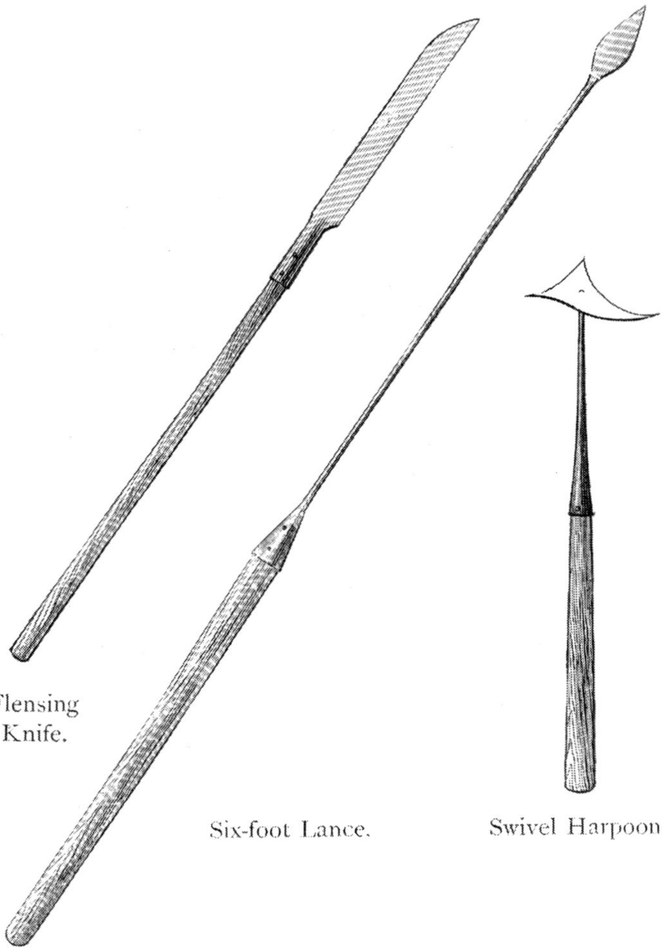

Flensing Knife.

Six-foot Lance.

Swivel Harpoon.

Chapter III.

THE SOUTH-WEST FISHING.

THURSDAY, *May* 15*th*.—In the forenoon the ice-blink, which is a light whitish tint along the horizon, proclaiming the immediate vicinity of ice, was plainly visible. Several large icebergs were passed, one of which was of enormous dimensions. We estimated its length to be at least a mile, and its height between two and three hundred feet. When first seen, I was under the impression it was land distant about seventy miles off, and it was some little time before I was convinced that it was really and truly an island of ice.

Shortly after noon we arrived at the edge of a large and compact stream of pack ice, and, seeing clear water beyond, the ship was pushed in. The ice, however, was found to be of a heavier description than was anticipated; in fact, as the captain observed, it was a mass of floating bergs, in which we were, for a time, hopelessly jammed.

Though there was a fresh breeze blowing at the time, and the ship had every stitch of canvas that it was possible to set, with the engines going ahead full speed, we were immovable for at least twenty-five or thirty minutes. From the deck, as far as the eye could reach, the ice was continuous, even the wake of the ship astern being closed up.

Eventually, by dint of screwing and boring, we succeeded in forcing our way through this stream of ice, and reached the clear water, or *polynià*, ahead, though not without experiencing a little rough handling,—collisions by which the ship was brought to a standstill,—sufficiently serious to damage any ordinarily built ship.

Shortly after four, our observant and ever-watchful captain espied three whales from the crow's nest. Instantly all was excitement; the main-topsail was backed, and four boats quickly lowered and in chase. No sooner had they left the ship than other whales were seen, and by five o'clock every boat in the ship was away and in pursuit. Though the whales were numerous, and the harpooneers used their utmost exertions, they were unable to approach sufficiently near to strike. It was, however, with the greatest reluctance, after five hours' ceaseless pulling, that the captain ordered the boats to be recalled. We must ascribe our want of success to-day either to the timidity of the whales, or the extreme clearness of the water.

During the absence of the boats, we on board were fully occupied in the multifarious duties connected with the working of the ship; and taking into consideration the few remaining hands left, we succeeded in tacking several times in a very creditable manner.

There seems no doubt that we have hit off the right spot, and all, though a little disappointed at the result of the day's exertions, confidently look fo-.ard to the prospect of plenty of work during the ensuing ten days.

It is a great thing to be the first ship on the ground, as the whales will not have been frightened by previous attempts at capture. The engines are not now used, as the slightest noise made by the screw, or in fact anything *under* water, would at once, to use a whaler's phrase, "scare the fish."

The whalers are very arbitrary in their phraseology, for although a whale is not a fish, with them nothing is a fish but the Greenland whale. For the future I shall adopt their ways, and call the mammal, the capture of which is our great object, a fish.

Our days are now very long, the night seldom being of more than three or four hours' duration; and the reflection of the sun, which is a very short distance below the horizon, is clearly visible to a very late and at a very early hour.

Friday, May 16th.—This has been rather an eventful and exciting day, though the result has not turned out as we could all have wished. A little before noon a fish was seen. Two boats were immediately lowered and sent in pursuit. We were at this time close to the edge of very heavy pack ice. After an anxious half-hour had passed, the whale was seen to rise close to the boat in which was the speksioneer. All was now breathless excitement, which was increased, if possible, when we saw the bow oar laid fore and aft, and the speksioneer rise to his gun. A puff of smoke, a moment of intense anxiety, relieved quickly by the captain's voice from the crow's nest calling, "A fall! a fall!" and the same cry borne along the water from the successful boat, and we knew that Davey Smith, the speksioneer, was fast. Immediately the upper deck was alive with men, all frantically shouting, "A fall! a fall!"[1] and rushing

[1] The cry, "A fall! a fall!" which is always called the instant a boat is fast to a fish, according to Scoresby is derived from the Dutch language. The word *val*, signifying to jump, &c., is considered expressive of the conduct of the sailors when manning

to the boats in readiness for an instant departure. All the men appeared to have gone suddenly distracted, and were tearing about the decks half dressed, with their boots in one hand and a bundle of clothes in the other. Four boats were ordered to be lowered, and despatched with instructions to spread in different directions, the more readily to strike another harpoon into the whale when it next rose to blow. Meanwhile the fast boat[1] had hoisted its jack,[2] to denote that it was fast, a second boat pulling up at once in order to bend on their own lines if necessary. The fish, on

---

the remaining boats for the purpose of proceeding to the "fast one." A more probable derivation is from the Dutch word *wall* (pronounced *vall*), a whale.

Some of the old whalers, however, are under the impression that when "a fall!" is cried, it implies that the men on board are to stand by the falls of the boats in readiness to lower. There are numerous old and quaint words and phrases made use of in the whaling trade, whose origin it would be difficult to trace.

[1] In the whale fishery, any boat which has a harpoon into a whale is denominated a "fast boat," all others "loose boats." So also with the whales: one that has been struck is called a "fast fish," others "loose fish." Should the line break and the harpoon still remain in the whale, it is a "loose fish," and may become the prize of any other ship, notwithstanding the harpoon, with the vessel's name engraved on it, being imbedded in its flesh.

[2] A fishing flag or jack is always displayed by every boat that is fast to a fish.

being struck, had, as anticipated and feared, made for the ice.

We in the ship, having steam at our command, immediately followed, steering for the pack, which we shortly entered; and then ensued a scene which almost baffles description. It was blowing a stiff breeze, and there was rather a heavy swell on at the time. The fast boat had been brought to a stop by a heavy floe, and the other boats, which had all entered the loose pack in pursuit of the fish, were being so severely handled by the ice, that for some time great fears were entertained for their safety. On several occasions they were nearly crushed between the floes, and were only saved by the promptitude of their crews, who, hastily jumping out, would haul them up on the ice. Some of these floes were drawing over twenty feet of water, which may give an idea of the ponderous weight of these floating masses of ice. After a little time had elapsed, and not without a great deal of difficulty, we succeeded in picking up our four loose boats, and then directed our attention to the fish, which had been observed to come up in a water space, but at such a distance that we were unable either to use a lance or to fire a second harpoon. Our first project was to transfer the lines from the fast boat to the ship, which was accordingly done, though not without much trouble, as the fish was still taking line, and at such a rate that the bows of the boat were drawn down to the water's edge, and the harpooneer was enveloped in smoke caused by the friction of the line round the bollard head. The whale had by this time run out ten lines (1,200 fathoms), equivalent to about a mile and a-quarter.

The captain was much afraid lest he should lose all these lines, as he had from the first foreseen the difficulty that must necessarily ensue in attempting to capture a whale amidst such heavy ice.

However, no one despaired, and "all hands" manned the line, resolved to bring the fish home, carry away the line, or draw the harpoon. Every one worked well and cheerily, knowing that £1,000 was at the other end of the line. It was with no little surprise and wonder that I witnessed, for the first time, the enormous power and strength of these leviathans of the deep. Not only were we being towed by the monster through the pack, but with such rapidity that we were frequently brought into violent contact with the heavy floes.

Things were beginning to look brighter; heavy strains had been brought on the line, and still everything held; the men were singing cheerily, and already counting up their oil money; more than half the line had been hauled in, and we were all confidently looking forward in a short time to behold our prize. Suddenly a more than usual strain came upon the line, a quick and sudden jerk, the line ran in easily, and we knew our fish had escaped. A bitter sense of disappointment seemed instantaneously to settle upon everybody. "She's gone!" was re-echoed through the ship; no more jokes were cracked, no longer was any singing heard, and the line was hauled in slowly and silently.

The fish had been struck with two harpoons—the gun and the hand; the latter, it appeared, had drawn, and in doing so its sharp edge had severed the fore gore to which the gun-harpoon was attached. This latter the unfortunate fish carried away embedded in its flesh, a painful and uncomfortable *souvenir*. The hand harpoon, which we recovered, was bent and twisted in a surprising manner. We had been fast to the fish over four hours.

The captain, though sorely tried, soon recovered his usual good temper and jovial spirits, consoling himself by saying that it might have been worse, as he might have also lost all his lines. We must hope for better luck next time. We succeeded in steaming out clear of the pack at about 6 P.M., and, putting the ship once more under canvas, "lay to" in a regular bight formed by the ice, which is regarded as a likely looking place for fish.

Tuesday, May 20th.—The last three days have been passed with the usual bustle and excitement attendant on seeing whales; boats have been lowered, but have always returned unsuccessful, after long and wearisome chases. Occasionally a boat would approach nearly to striking distance, but only to be disappointed by the fish escaping under the ice just as the harpooneer was rising to his gun. These little failures are all most annoying and vexatious; it is very tantalizing getting so many chances, and yet always being unsuccessful in our attempts to secure a prize. Our men are very superstitious, and attribute our ill-luck to various causes. One day it is put down to a *comb* which is universally used by all in the cabin, and which, in consequence, nearly fell a victim to their superstition; another day it is to a small pig we have on board, and which, I have no doubt, if we do not get a fish, will soon be offered as a sacrifice at the shrine of Dame Fortune. I trust they will not impute their ill-luck to the fact of my being on board, imagining that a naval officer is as unlucky on board a whaler as some sailors fancy bishops to be. Other whalers have joined us, and yesterday, to our mortification, one of them, the "Narwhal," succeeded in capturing a fine fish under our very eyes.

The whale had been first seen by one of

our men, and two boats lowered and despatched in pursuit, the other ships also sending their boats. These were all spread out, occupying a distance of at least a mile, when, as ill-luck would have it (as far as we were concerned), the whale rose close to one of the "Narwhal's" boats, which immediately struck the fish, and up went the flag as a signal that they were fast, on perceiving which the "Narwhal" hoisted her jack,[1] and sent the remainder of her boats away to assist in killing the whale, the boats belonging to the other ships being recalled. Meanwhile the unfortunate fish, which had gone down to a considerable depth on being struck, came up to blow, when another harpoon was fired into it, and another flag was hoisted, announcing that two boats were fast, and so this continued until there were five boats fast, leaving three others free to come up and despatch the whale with their lances. By that time, however, we were at too great a distance for me to observe any of the details. The whale ships always have their boats painted different colours, the more readily to distinguish them from other boats when fishing in

[1] When ships are fishing in company, it is usual for the vessel whose boat succeeds in getting fast to a fish, to display the fishing jack from the mizen top-gallant mast head, which is kept flying until the fish is killed. Each ship has a different device or pattern for her fishing flag.

company. Ours are painted blue, white, blue, longitudinally, with black gunwales; and our fishing flag is white, with a blue five-pointed star. We keep dodging about amongst the ice, poking into water-holes, always keeping a bright look-out for fish. Several white whales and narwhals keep sporting about the ship, but these are seldom

Iceberg.

molested when there is a chance of obtaining larger and more profitable game. Numerous icebergs are in sight, some of enormous size. Yesterday we passed to leeward of one whose circumference must have been over two miles.

We have seen land to the westward, which we suppose to be Resolution Isle, distant about twenty-five or thirty miles. We attempted to obtain soundings, but got no bottom with two hundred fathoms of line out. I brought into use on this occasion the Casella thermometer with which the Hydrographer had kindly supplied me, and ascertained that the temperature at that depth was 30°, whilst at the surface it was 34° the air at the same time being 39° This result must at once put a stop to any idea that may be entertained of a warm current flowing up Davis' Straits.

This evening we forced our way through several wide and apparently impervious streams of ice, which kept the men incessantly at work, either throwing everything aback to avoid a collision with an unusually large and heavy floe, or else backing the main-topsail only, to deaden the ship's way.

It is most interesting to watch the large masses of ice crack, and in some cases crumble up, as the ship comes into violent contact with them, and to observe the masterly and skilful manner in which the ship is handled by the captain, or officer stationed in the crow's nest.

Thursday, May 22nd.—Our ill-luck continues to remain with us. Every day we see fish, and the boats are sent away in chase; but the fish show sufficient sense to make themselves scarce on the approach of their would-be captors, and the boats consequently always return unsuccessful. When the ship is not in motion, one or two boats are always kept manned by the watch on deck, which lay off the edge of the ice, or where ordered, in readiness to pursue a fish the instant it is seen. This is called keeping a boat "on bran." We are now, to our great delight, entirely by ourselves, the remainder of the fleet having sought a fishing ground further to the westward. Yesterday I observed the wonderful distortion caused by refraction. The "Intrepid," which was hull down on the horizon, probably about eight miles off, was drawn out into an elongated and most fantastic shape. Her funnel seemed to rise as high as her main-top-gallant mast head, whilst her sails assumed the most quaint and wonderful forms. Icebergs, also in the distance, seemed to have no connexion with the water, but appeared suspended as it were in mid air.

Life on board a whaler, though occasionally enlivened by a little excitement, is after all rather weary and monotonous when confined to an area of a few miles in extent to cruise about in, especially when one's thoughts are always on the unexplored regions of the far north. At present our captain's great idea, should the whales be scarce in Baffin's Bay, is to push up Prince

Regent Inlet into the Gulf of Boothia, that in his opinion being the great resort and breeding-place of the mysticete. Little is known of that interesting locality, no whaler having as yet had the enterprise to navigate its waters. The whale ships, as a rule, unless commanded by some dashing and energetic man, invariably, and year after year, fish upon the old ground, and seldom leave the beaten track.

Eglinton Inlet—a favourite resort of the whalers in Davis' Straits.

Chapter IV

" FLINCHING" AND "MAKING OFF."

AT length, after several near "shaves," I am enabled to chronicle the capture of a whale, and to describe the subsequent operations connected with the cutting up or "flinching" of the fish.

At four o'clock this morning a whale was seen, and a couple of boats lowered and sent in pursuit, but the chase was apparently abandoned shortly after five, the captain looking into my cabin, and in forcible terms expressed his opinion, that notwithstanding his having "turned his horseshoe" yesterday, we were going to have no luck this cruise. He then went to bed. At half-past seven our ears were assailed by that cry, which is such sweet music to all on board a whaler, "A fall! a fall!" accompanied by the usual commotion on deck. Hurriedly putting my coat on, being in the

THE "ARCTIC," WITH BOATS FAST TO A "FISH."

Page 50.

act of dressing at the time, I hastened up and found that the speksioneer was again the lucky man, having struck a fine fish, which was rapidly towing him towards the edge of the pack ice. Our boats were soon in the water, and in less than twenty minutes the unfortunate fish had six harpoons buried in its body. In the mean time

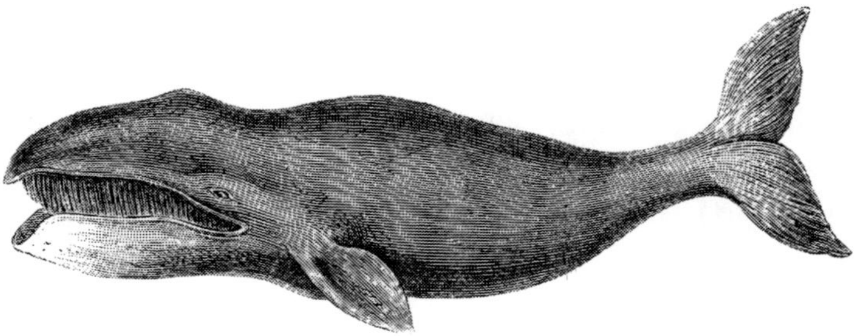

Whale (*Balœna Mysticetus*).

I had retired below to complete my toilet, the scanty manner in which I was attired being ill adapted for remaining on deck with the temperature four degrees below the freezing point. On my return to the upper deck, the loose boats were observed alongside their prey in the act of administering with lances the *coup de grâce*, on receiving which the huge monster turned over on its back and expired. The flags in the fast boats were then struck amidst the cheers of the men both in the boats and on board the ship. The loose boats were at once recalled to the ship, their crews, after the boats had been hoisted up, being busily employed in making the necessary preparations for flinching and taking in the blubber. The crews of the fast boats were engaged in hauling in and coiling away their lines, which on the death of the whale are always cut at the splice of the fore gore. It is usually the duty of the crew of the *first* fast boat to prepare the whale for coming alongside. The preparations consist in cutting a hole through each fin, for the purpose of receiving the fin-tow, and lashing them together across the belly of the fish. This is done in order that they may offer no obstruction in the water whilst the fish is being towed alongside. The tail is then roused up to the bows of the boat, and in this way the fish is brought to the port side of the ship, and there secured in the following manner. The fish is always brought alongside with its tail forward abreast of the fore chains; it is then secured by means of a tackle from the fore rigging, which is hooked to a strop round the small end of the tail (where it is united to the back of the fish), and by a stout rope, which is called the "rump rope." A similar purchase is hooked from the main rigging to a strop rove through a hole cut in the extremity of the under jaw, which is called the "nose tackle." The right fin of the fish (which is next the ship's side, the whale being on its back) is dragged taut up and secured by a chain or rope to the upper deck, the bulwarks of the vessel on the port side being unshipped. Between the fore-mast and main-mast is a stout wire rope, called the "blubber guy," having four large single blocks strapped to it, through which are rove the fore and main spek tackles (five-inch rope). The former is usually worked by the steam winch, and the latter by a hand winch near the main-mast. These tackles are used for hoisting on board the large layers of blubber, some between one and two tons in weight, as they are cut off.

From the main-mast head is a heavy purchase called the "kent"[1] or "cant" tackle, which is used to turn the fish over as it is being flinched. It consists of a treble and a double block, having a seven-inch fall. Everything being in readiness, the crew are turned up, and having been primed with a glass of grog all round, commence the operation of flinching. Two boats, called "mollie boats," attend upon those engaged in cutting up, and are kept alongside the fish by a couple of hands in each boat, who are named "mollie boys," the lines having been carefully covered over with boards to prevent injury. The captain, from the

---

[1] *Kent* derived from the Dutch signifying to "cant" or "turn."

port main rigging, superintends the whole process; the mate in the gangway acting under his orders. The remaining seven harpooneers, under the guidance of the speksioneer, are on the whale, and with their blubber spades and knives separate the blubber from the carcase in long strips, which are hoisted in, as before mentioned, by the fore and main spek tackles. Previous to this, however, a strip of blubber, from two to three feet in width, is cut from the neck, just abaft the inside fin. This is called the "cant." A large hole is

Blubber Spade.

then cut in this band of blubber, through which is passed the strap of the cant purchase, and secured there by a wooden toggle or fid being passed through. By means of this purchase, brought to the windlass, the fish is turned over as required. Each harpooneer has iron spikes, called "spurs," strapped on to his boots, to avoid the possibility of slipping off the fish.

The belly is the first part of the whale that is operated upon. After the blubber from this part has been completely taken off and the right fin removed, the fish is canted on to its side by means of the large tackle, and the blubber from the opposite side is similarly stripped. The whalebone is then detached, special bone gear being used for this purpose, and the lips hoisted in, and so on until all that is valuable has been cut off and taken on board. The tail is then separated from the carcase, or "kreng," as it is called, which latter being released, disappears with a plunge, the noise of which is drowned only by the cheers of the men, discolouring the water with a sanguine hue for some distance. The duties of the boat-steerers during this operation are to cut up the large strips of blubber as they are received

Pick-haak.

on deck into pieces about two feet square, with long knives. These pieces are seized by the line managers, armed with "pickies" or "pick-haaks,"[1] and transported below through a small hole in the main hatchway. Below they are received by the "skeeman," and another man denominated a "king," by whom they are stowed temporarily between decks, until such time as an opportunity may offer for performing the final operation of "making-off."

---

[1] A Dutch word.

The whalebone, on being received on deck, is split up into portions, each containing from nine to sixteen blades, by means of large iron wedges, and these are again divided into pieces of three or four blades, when what is called the gum, which connects them together, is removed. There are between three and four hundred of these blades in each side of the head.

The tail of the whale is cut up into blocks, which are used during the process of "making off," forming excellent blocks, on which the blubber is chopped up into small pieces, thereby preserving the edge of their instruments. It is customary for the boat-steerer of the first fast boat to have the choice of blocks, after which selection each man marks his own by cutting his initials, or otherwise distinguishing it.

After the operation of flinching is concluded, the upper deck, as may be imagined, is in a very filthy state, and so slippery that unless great care and caution are exercised whilst walking, a fall is inevitable. A little sawdust, however, sprinkled over the worse parts, makes it a safer promenade.

Our men were remarkably quick in flinching this our first whale, the "kreng" sinking, amidst the frantic cheers of all hands, exactly two hours after the operation had commenced.

The crews of these ships are always ready to

cheer on the slightest provocation. It is customary to give three cheers, both when the fish is killed and when the last of it is taken on board. This capture already seems to have had a wonderful effect upon every one on board; all are now in high spirits, predicting a successful cruise, and a return to Dundee with a full ship.

The men attribute their good fortune to-day to the fact of their having last night burnt in effigy two of the crew, who are supposed to bring ill-luck, having had the bad fortune to serve of late years in ships which have returned "clean," or after very poor voyages. Our prize to-day was not what is considered a large one, although regarded as a fair average-sized fish. It will probably yield about thirteen tons of oil; its bone was 9 ft. 6 in. long; altogether about the value of £800; its length was between forty and fifty feet. The six harpoons, which were taken out of the whale during the process of flinching, were twisted and bent into most extraordinary shapes, thereby fully demonstrating the necessity of their being made of soft and pliable iron, so as to yield without snapping, and at the same time to resist the enormous power exerted by the fish in its endeavour to escape from its relentless enemies.

Whilst engaged in securing the fish alongside, the steward, who has a remarkably quick eye, espied a bear upon an iceberg about two hundred yards from the ship. Running down for my rifle, I jumped into the dingy, and with the doctor and a couple of hands pulled in the direction it had been seen. On rounding the berg we observed Master Bruin, who trotted down towards us, apparently to make a closer inspection, and satisfy

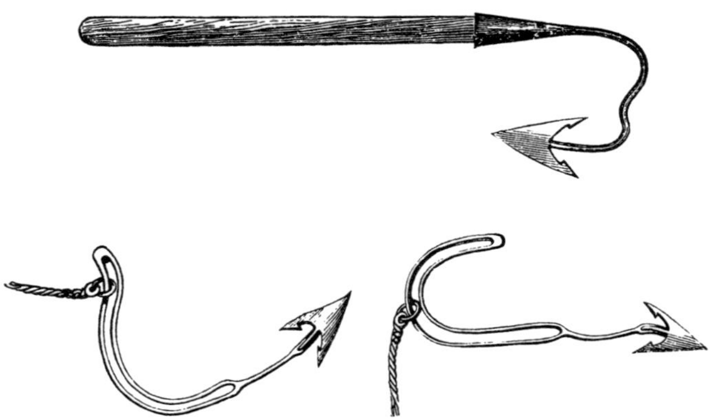

Twisted Harpoons.

his own curiosity. A bullet from my rifle entering his shoulder caused him to beat a rapid retreat, endeavouring to effect it by taking to the water, and avoiding us by swimming and diving. I fired a second shot at about fifty yards, which, passing through his skull, killed him instantaneously. He was a fine young bear, measuring close upon seven feet. The skins of bears, after being flinched, are treated in the same way as are those of seals, a large quantity of blubber being taken off them.

We are now in a large water-hole, entirely surrounded by ice. The captain does not anticipate much trouble in getting clear, though we shall probably have to bore through over fifty miles of rather heavy ice!

Saturday, May 24th.—This being a fine calm day, and no whales having been observed, advantage was taken to perform the operation of "making off." To carry out this duty, in which "all hands" are engaged, it is necessary to select a quiet day, to enable the men to complete the entire operation without being disturbed.

The blubber is first hoisted on deck by means of the main winch, worked by the firemen. It is then seized by two men on each side of the deck, who, with their pickies, drag it to two men stationed on each side (generally harpooneers), whose duty it is to cut it up in pieces about twelve or sixteen pounds weight, and who remove from it all kreng and other extraneous matter. These men are called "krengers." The blubber is then thrown forward to the remaining harpooneers, who are stationed on each side of the deck near a "clash," which is an iron stanchion firmly fixed into a socket in the deck, standing about three feet high, and having five iron spikes on the top.

Each harpooneer, or "skinner," as he is called whilst so employed, has an assistant, who is called a "clasher," who picks up the pieces of blubber having skin on with a pair of clash-hooks, and places it on the top of the clash. The skin is then separated from the blubber by the skinner, armed with a long knife. The blubber is then deposited in a heap, which is called the "bank," directly in front of the "spek trough," which latter is a large oblong trough, about eighteen feet long, and two feet in width and breadth,

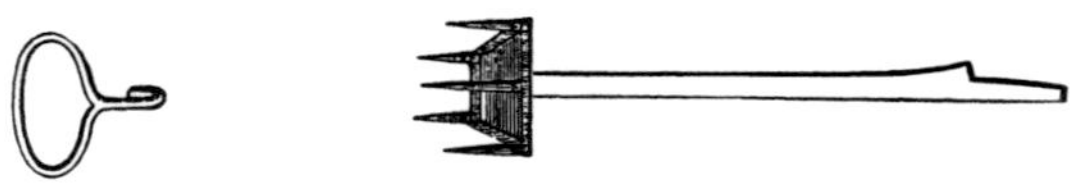

Clash and Clash-hooks.

which is placed immediately above the hatchway, through which the blubber is to be passed down. A hole about a foot square is cut in the centre of this trough, to which is fitted a long canvas shoot or hose, called a "lull," the end of which is pointed into the tank receiving the blubber. The lid of the trough is turned back, and is supported underneath by chocks, so as to form a table about three feet high, on which are placed the blocks cut from the whale's tail. Behind these blocks are stationed the boat-steerers, armed with choppers, whose province it is to chop up the pieces into small portions, after they have passed through the hands of the skinners. They are then thrown into the spek trough, passed down through the lull, and so into the tanks.

The skeeman and king superintend its transit below, and shift the lull from tank to tank as they become full. One hand is stationed at the grindstone for the purpose of sharpening the different instruments as required, the remainder of the men, armed with pickies, passing the blubber to and from the various operators.

This work tends to make the ship in a more filthy and greasy state than the operation of flinching, although there is nothing absolutely repugnant or disgusting in witnessing the process. Indeed, the upper deck, during the time the work is at its height, presents a most animated and busy scene. Forward, standing in a line across the forecastle behind their blocks, are the boat-steerers, with their continual and ceaseless chopping; in front of them are men busily employed with pickies, transferring the blubber (which has rather the appearance of huge lumps of cheese) from the deck to the spek trough; whilst on each side are the skinners, with their assistants, engaged in their individual labours. All is life and activity, every one in a good humour, and working with that cheerfulness and energy which are the result of a contented and happy disposition.

The whale skin, which is generally thrown overboard, is on board the "Arctic" carefully collected and placed in a tank, by order of our kind and thoughtful captain. On the arrival of the ship at Lievely, it will be by him distributed amongst the Esquimaux inhabitants of that settlement, by whom it is much relished and appreciated as an article of food. It is also regarded as an excellent anti-scorbutic.

During the proceedings of yesterday and to-day the ship was literally surrounded by hundreds of screaming and greedy fulmar petrels (called by the whalers "mollies"), clamouring and fighting over the numerous pieces of kreng and blubber that floated alongside and astern of the ship. These greedy birds are at times so voracious as frequently to alight on the whale to pick up some choice and delicate morsel, whilst the men are actually employed on the fish in the duties of flinching.

We are now amongst loose pack ice, our large water-hole of yesterday being filled with numberless floes.

From the crow's nest there is no open water visible, nothing but ice as far as the eye can reach. To one having such limited experience in these matters as myself, we appear to be hopelessly surrounded and completely jammed in by the ice, and it seems almost impossible, without a change of wind, that we can succeed in extricating ourselves without recourse to cutting or blasting. Even the captain anticipates a difficult and laborious passage through.

Fulmar Petrels, or "Mollies"

CHAPTER V

NAVIGATION OF DAVIS' STRAITS.

SUNDAY, *May 25th.*—This morning the ice closed in rapidly upon us, and from 6 A. M. until 9 P. M., when we emerged into tolerably open water, we have been boring and pushing our way under steam and sail through interminable fields of ice. At times we were immovably fixed between heavy floes, making no headway whatever, then again we were rushing through young and brashy ice, splitting and driving the fragments before and on each side of us. Occasionally we came into contact with an unusually heavy piece with such violence as to cause the ship to *recoil* several yards, when she again gathers way and again charges her almost impenetrable foe.

As far as the eye could see was nothing but ice, with here and there narrow streams or leads of water; altogether a most desolate and cheerless scene, the monotony of which was only relieved by the numberless icebergs dotted around the horizon, which imagination could almost fancy to be steep cliffs and head-lands surmounted in some instances by castles and towers.

Iceberg.

To enhance the general wintry aspect, snow was falling heavily during the forenoon, the deck and rigging being completely covered; the captain himself with his beard and whiskers so encrusted as to resemble, with his portly person and jovial face, the drawings of Father Christmas. We were nearly sixteen hours before we succeeded in extricating ourselves from our icy prison, during which time we must have bored through at least fifty miles of pack ice. Had we been suddenly deprived of our steam-power we should have been in a most unpleasant and precarious predicament. A sailing-vessel would have occupied as many days as we did hours to get through. The general course that we steered throughout the day was about due east (true).

This part of Davis' Straits, commonly known to the whalers as the S. W fishing-ground, has generally had the unenviable reputation of being a dangerous neighbourhood, and one which it is most desirable for a ship to get quickly clear of. My own experience of it, during the last ten days, though we have had unusually fine weather, impressed me strongly as a locality which should, as much as possible, be avoided by vessels without steam-power.

From the time these waters were first navigated, ships have invariably encountered great dangers and the crews much hardships in this dreary vicinity.

During Sir Martin Frobisher's third voyage to these regions in the year 1578 it is related that the barque "Dennis" of 100 tons "received such a blow from a rock of yce, that she sunk down therewith in sight of the whole fleet;" after which fearful catastrophe "a sudden terrible tempest arose from the south-east, the ships weathering which became encompassed on every side by yce, having left much behind them, thorow which they had passed, and finding more before them thorow which they could not passe. Some of the ships, where they could find a place more cleare of yce, and get a little berth of sea roome, did take in their sayles, and there lay adrift; other some fastened and moored anker upon a great island of yce; and again some were so fast shut up, and compassed in amongst an infinite number of great countreys and islands of yce, that they were fain to submit themselves and their ships to the mercy of the unmerciful yce, and strengthened the sides of their ships with junk of cables, beds, masts, planks, and such like, which being hanged overboard, on the sides of their ships, might better defend them from the outrageous sway and strokes of the said yce." One of their methods for boring through fields of pack ice was decidedly primitive, but it was then considered novel and ingenious, and appears to have been successful. We are told, during this same voyage of Sir Martin Frobisher, that the "Judith," having parted company from her consorts, was beset by the ice, "and when that by heaving of the billowe, they were therewith like to be brused in peces, they used to make the ship fast to the most firme and broad pece of yce they could find, and binding her nose fast thereunto would fill all their sayles; whereon, the wind, having great power, would force forward the ship, so the ship bearing before her the yce; and so one yce, driving forward another, should at length get scope and sea roome. Having by this means, at length put their enemies to flight, they occupyed the cleare space for a prettye season, among sundry mountains and Alpes of yce." (What would our skilful whaling captains of the present time think of this mode of forcing a passage through the ice?)

Turning to later voyages, Parry, in 1821, was closely beset by the ice in this same locality, consisting of loose masses of broken floes, amongst which his two ships the "Hecla" and "Fury" drifted about for fourteen days before they were able to extricate themselves. He speaks of the scene as being "indescribably dreary and disagreeable."

The gallant Sir John Franklin also bears testimony to the dangers to be apprehended by navigators in these waters, for whilst on his way out in 1819, to undertake in conjunction with Dr. Richardson and Sir George Back (then Mr. Back) that memorable and perilous journey over land to the shores of the Polar Sea, he alludes to "that dangerous, and by all abhorred, island Resolution, in the mouth of Hudson's Straits, near the rocky shores of which, usually beset with heavy ice, fogs, and irregular currents, the vessel narrowly escaped shipwreck;" and lastly that distinguished officer, Admiral Sir George Back, himself, when in command of H.M.S. "Terror" in 1836, previous to those extraordinary and unprecedented perils for which that disastrous voyage is memorable, bears record to the general abhorrence and fear with which this part of Davis' Straits is regarded, in the following words, "on approaching that universally detested Resolution Island, with its dense fogs, and its whirlpools, tossing about masses of ice, sweeping the ship among them and rendering her utterly unmanageable." It will thus be seen that all who have had experience in these regions are unanimous in their opinions regarding this neighbourhood.

Many of the bergs which we passed to-day were of great magnitude, and are indeed "large countreys and islands of yce." We saw none, however, that could compare in size with that one of gigantic dimensions seen by Master John Davis (after whom the Straits are named), who, on the 17th of July, 1586, "fell in with an enormous mass of ice, having all the characteristics of land." He very wisely declines entering into particulars as to its size and height lest, as he says, "he should not be believed." Some notion of its magnitude, however, may be formed from the fact that he sailed along it till the 30th; and that while in its vicinity, "the cold was extreme, the shrouds, ropes, and sails being frozen, while a dense fog loaded the air."

Tuesday, May 27th.—During the last two days we have had a succession of very heavy snow showers, and although the men were kept constantly employed in shovelling and clearing the snow away, it was frequently laying a foot deep on the upper deck, and the masts, yards, rigging, and ropes were completely covered.

The general set of the ice, and consequently of the ship, has been to the southward. During the last forty-eight hours it has been very strong, at least seventy miles for the two days. To-day we are struggling against a strong northerly wind; however, "out of evil comes good," and although we are now labouring under the disadvantage of a head wind and an adverse current, still we hope to derive some benefit from it on our arrival in Melville Bay, as it will cause the ice there to be loosened, thereby affording us a safer and quicker passage through; but we must not anticipate, as a strong southerly wind will again pack it tight and close.

We are making short tacks under reefed topsails

and foresail, working along the edge of the pack, by which, in a great measure, we keep comparatively smooth water. The loose floes and bergs are all drifting rapidly to the southward and westward towards the main pack, and we congratulate ourselves in having so easily escaped the danger of being beset in a frozen prison.

We have quite given up the idea of capturing any more fish in this neighbourhood, the whales having all left for the north, where we are now doing our best to follow them. When the wind subsides steam will be raised and a course shaped for Disco.

Thursday, May 29th.—We seem destined to have nothing but head winds, accompanied, as to-day, by a nasty chopping sea, having the effect of making the old ship "pile drive" considerably, so much so that it is with extreme difficulty I am able to write, holding the inkstand with my disengaged hand, everything having a tendency to slide away to leeward.

Occasionally we come into contact with a stray piece of ice, unobserved in consequence of its deep flotation, which makes everything dance on the table, and even spills the ink out of the stand. Last night we passed a remarkable looking berg, which at first I mistook, and very naturally, for a rock. It was of a dark brownish colour, rounded on one side, over which the sea was dashing, and rising to a height of about twelve or fourteen feet out of the water. It was only when we approached and were able to make a closer inspection that we were convinced it was ice, discoloured probably by the muddy water of some mountain stream. One large berg which we passed shortly after had two clear and distinct red lines, which met about half-way towards its base in the form of a wedge, having its apex at the bottom. This must also have been caused by alluvial deposit. The number of icebergs seen to-day is almost incredible, fully seven or eight hundred, and some of great magnitude. During the afternoon a little snow bunting was caught, which, however, did not survive its capture long. It had evidently come from some distance, and was quite exhausted.

Our captain always carries about with him what he calls a "lucky penny," one of those huge coins in circulation in the reign of George III. With this, from the first thing in the morning until the last thing at night, whether on deck or below, he is always anxious to toss with the doctor for the best of five successive guesses. It is most amusing to watch the cunning manner in which our worthy skipper puts his coin down, and the delight beaming on his jolly countenance when he succeeds in winning. As a rule they toss for nothing but their own amusement, but in the evening it is generally for who shall fetch the plates, &c., for our supper, the water for our toddy, or the glasses for our grog. In the evening they amuse themselves with cards, playing rather a noisy game, peculiar, I presume, to whalers, called "Scratch the ten," whose particular feature appears to be a grand fight as to which shall obtain possession of the ten of trumps.

I remained up, watching the sun as it slowly disappeared below the N. W horizon, sinking gradually and majestically. It was past ten before its upper limb was out of sight, but such a short distance was it below the horizon that it was really only a matter of form that we had a candle lighted in the cabin, and when I retired to bed at midnight there was sufficient light in my little cabin to read by.

Friday, May 30th.—The wind having gone down, the engines are once more brought into play, and we are again propelled by the "brass topsail." We are still hugging the ice, and have passed through various deep bights along the edge of the main pack, and through several broad streams of ice. The late northerly winds having caused a heavy swell to arise, the heaving, rolling pack through which we steamed had a most indescribable appearance, bringing vividly to my imagination the dangers encountered by the little "Fox" whilst under the command of Sir Leopold McClintock, when she successfully steamed out of the rolling pack after having undergone the vicissitudes of an Arctic winter. One must actually experience the dangers of Arctic navigation before they can be realized. Several of the floes that we passed to-day were piled up with masses of hummocky ice, formed in solid square and cubic blocks, resting in an uneven and confused manner one on the other, and presenting in places, varied brilliant tints of blue.

These hummocks are formed by the action of one floe against another, pressing and squeezing off by their ponderous weight these fragments which rest on the floe where they have fallen.

It is most interesting to watch from the forecastle the ship battling, as it were, with the ice, one moment striking a floe stem on, causing the ship to come to a dead stop, at another making a cannon from one piece and striking a second on her opposite bow, which will rebound with such force as to make the ship's head swerve from four to six points ; again, there are other pieces which are pressed down by the weight of the ship's bow, and which, when released of the pressure, rise rapidly and suddenly in most unexpected directions. Great care must be taken to fend off these latter pieces by means of long poles, as they spring up in the water to such a height, and with such velocity, that they have occasionally been known to rise under one of the boats, completely smashing it. Several large icebergs were passed, one, the height of which we estimated at over four hundred feet. By way of making us feel easy and comfortable amongst so many of these huge masses of ice, the captain beguiles the time by recounting to us a few of his own personal experiences during his long and arduous services in these regions. Some of these I will attempt to relate.

Whilst serving as harpooneer on board a whaler (in much about the same position in Davis' Straits as we are to-day), his ship was running under reefed topsails and foresail at the rate of ten knots an hour before a strong southerly gale, snow falling heavily, and the weather, to use a seaman's expression, as thick as pea-soup. The men were employed in getting the boats in-board, and securing them from the heavy and angry white-topped waves that came curling up astern and on each quarter, apparently eager to overtake and swallow up the little vessel that was flying in mad career along their crests. Suddenly a huge mountain of ice was seen ahead. To have struck it would have been total destruction to the little craft, and death in all its horrors to those on board. "Hard a-port" was yelled in agonized tones from the look-out on the forecastle, "hard a-port" was echoed along the deck, and answered quickly and smartly by the ever-watchful and ready helmsman. To answer and to act was the work of a moment, a breathless suspense for all on board, and amidst the sighing of the wind, the creaking of the masts, and the rattling of the blocks and cordage, accompanied by a far more fearful and dreadful sound, namely, that of the sea dashing over the berg, the little ship, obedient to her helm, flew up to the wind, and as she did so, the eddy and the back-water from the lumbering mass of ice nearly threw her upon her beam ends to *windward.* She quickly righted, and as the wind again caught her sails, her lee yard-arms scraped along a steep and precipitous berg, causing fragments to break off, which actually fell upon the ship's deck. In another moment they were clear—and saved! the enemy with whom they had had so close a death-struggle lost to sight in the thick snow and foam astern.

In addition to these perils, which must necessarily be encountered by all who navigate these waters, the whalers have also to undergo the dangers which must constantly be met with whilst engaged in their vocation. As the class of accidents of which the following is a specimen is, I am sorry to say, not of unfrequent occurrence, I will relate it. A whale had been struck on the east side of Baffin's Bay, but the line having carried away, it succeeded in effecting its escape, carrying, however, the not very pleasant souvenirs of two harpoons, which remained embedded in its flesh. By a curious coincidence, the same fish was fallen in with, some days subsequently, by the same ship, on the *west* side of the bay. Boats were immediately despatched for the purpose of securing their prize; but this was not easily accomplished. On the approach of the first boat the harpooneer fired and got fast, and was in the act of putting in the hand harpoon, when the fish, with a convulsive effort, struck the boat a blow with its mighty and ponderous tail, smashing it into a thousand pieces, and precipitating the crew into the water. The unfortunate harpooneer was no more seen, having probably been entangled by the line and taken down. The remainder of the crew, with the exception of one, were picked up by the other boats; the one exception had succeeded in swimming to a piece of ice, but in consequence of exhaustion and numbness caused by excessive cold, was unable to raise himself on it. He would shortly have perished had not one of the men, mistaking him for a *seal*, pointed him out to his companions. He was at once taken on board and restoratives administered, but it was many days before he recovered from the effects of his cold bath. When picked up his clothes were frozen hard on his body. The whale, the cause of all these misfortunes, was eventually killed, when the harpoons with which he had been struck on the east side were recovered.

Some accidents occur in whaling which savour more of the ridiculous. One man, though a good and successful harpooneer, could never be prevailed upon to approach a fast fish for the purpose of killing it, always exhibiting great timidity when ordered upon this service. On one occasion, nothing would induce this individual to pull alongside the whale in his boat and administer the death-thrust. He was, in consequence, upbraided by the other harpooneers for his cowardice. "What," said the second mate, "are you afeered for the fish?"—"No," answered the timid harpooneer, "but I am afeered for myself." And on another occasion, the same individual, whilst pulling towards a fish, was hailed by the captain from the crow's nest, and told "to keep off her eye," meaning to pull up behind the whale, so as to be unobserved, "Aye, aye," said our friend, "but I'll keep off the beggar's tail;" and nothing would induce him to go near that dangerous appendage of their formidable prey.

Listening to these and other stories of the whale fishery, and gaining an insight in the mysteries of ice navigation, the days pass quickly and pleasantly; seldom one passing without some quaint remark or odd proceeding on the part of some of my messmates. Our meals afford many opportunities for their witticisms; and if I request that I may be served with a piece of beef not quite a pound in weight, it is immediately remarked that the captain, as I am always called, is fond of "Waxhallers."[1] At dinner to-day we indulged in the luxury of green peas, very large and very hard; but I found that the effort of picking them off the plate with a "two-pronged steel fork" (the only kind we have on board) required so much patience and dexterity, and was withal of such a tantalizing nature, that it hardly compensated for the exertion required to ensure success.

Saturday, May 31st.—To-day we crossed the Arctic circle. Formerly it was customary on board whale ships to perform a ceremony somewhat similar to that still in vogue on board most men-of-war and other ships when crossing the equator. Since the introduction of steamers into this trade, the practice, like many others, has gone completely out of fashion.

In the forenoon we observed and communicated with the "Erik," bound, like ourselves, for

[1] Vauxhall Gardens were at one time famous for the extremely thin slices of ham and beef which used to be supplied to those requiring refreshments.

Lievely, as the Danish settlement on the south end of the island of Disco is called.

As this ship had left Dundee three or four days subsequent to our departure, we had the pleasure, through the kindness of her commander, of receiving letters from England. We also heard that, with the exception of ourselves, the "Narwhal" was the only vessel that had succeeded in getting any fish, she having been fortunate enough to obtain two.

Our ship has assumed quite a different appearance, the crew for the last two days having been actively employed in scrubbing the paint work and cleaning the upper deck generally.

The contrast is marvellous!

Iceberg.

CHAPTER VI.

DISCO.

SUNDAY, *June 1st*, saw us hammering away, under double-reefed topsails, against a strong northerly gale, making the ship in consequence very lively and very uncomfortable. By noon, however, it veered round to the westward, enabling us to lay a course for Disco.

The forenoon was rendered remarkable by the enormous number of icebergs, all of large dimensions, that we passed, most of them aground, off Rifkol. The captain estimated the number seen in four hours at about three thousand, and I should really be afraid to say whether he was under or over the mark. We passed within a very short distance of some of these huge islands of ice, their crests towering considerably over our mastheads, and against the sides of which the sea was dashing with uncontrolled fury. Many of these bergs were fully a mile in circumference and from three to four hundred feet in height. At half-past five we passed the Whale Fish Islands, so named from the supposed resemblance that they bear to the head and back of enormous whales lying on the surface of the water. They are also called the Kron-Prins Islands.

Formerly a settlement was established on one of these; but it has, I believe, of late years been abandoned.

The high land of Disco, which is a little over twenty miles from these islands, rose rugged and bleak, the summit of the hills being covered with a deep and impervious mantle of snow. Shortly after 7 P.M., passing close to the rocks on the south side, we shortened sail and steamed into the snug harbour of Lievely, anchoring off the picturesque little settlement where the Chief Inspector of North Greenland resides, an officer holding his commission direct from the King of Denmark. On the beach, on the north side as we entered, were lying the remains of the English whaler "Wildfire," which had been run on shore some years previously, after having sustained severe "nips" in Baffin's Bay. She is now a complete wreck. It was not long before several boats full of Esquimaux came alongside, all

DANISH SETTLEMENT OF LIEVELY (DISCO).

Page 82.

anxious to *trock*, or barter. Slippers and tobacco-pouches made of sealskin were the principal articles brought off, for which were sought in exchange powder, shot, coffee, shirts, and trowsers.

At about eight o'clock the captain, doctor, and myself went on shore, and were received on landing (an honour I little expected) by his Excellency the Inspector, attired in a uniform coat with shoulder-straps, all very gorgeous; but his extremities were encased in a pair of *sealskin* trowsers, affording a curious contrast.

After the ceremony of introduction had been gone through, he informed me that he had received notice of my intended visit from the captain of the "Esquimaux," who had called in some few days previously. We were then conducted to his house, and introduced to his wife, Mrs. Smith, and her sister. After spending a pleasant and quiet evening, we bade our hospitable entertainers good night, and after visiting a couple of Esquimaux habitations, which did not favourably impress me with a great idea of either comfort or cleanliness, returned on board at half-past twelve, the sun at this time shining brightly, and making it appear almost unnatural to think of going to bed; but exhausted nature must be attended to. From Mr. Smith we received the first intelligence of the safety of a portion of the crew of the U. S.

exploring ship "Polaris," who under Tyson had made that extraordinary and perilous drift on the ice of upwards of a thousand miles. The news had been brought from England by the "Esquimaux," which ship had left Dundee ten days after our departure.

Monday, June 2nd.—The "Erik" arrived early this morning, having been delayed outside in consequence of the strong wind. I went on shore in the morning with my artificial horizon, to try and get sights, but a heavy snow-storm coming on, effectually obscured the sun and precluded all observations. In the afternoon, accompanied by his Excellency, we walked over the settlement, and called upon Mr. Lawson, the Governor, who is a Dane, and subordinate to the Inspector. In the evening an Esquimaux dance was got up for our special edification, to which we all adjourned at about ten o'clock. It was held in a store-house, the casks and barrels having been placed outside for the purpose. The dimensions of this extemporized ball-room were about 18 feet long by 12 broad and 6 high! Into it were crammed our party, consisting of nine, about thirty Esquimaux, and about twice that number of men from the two whalers. The space left for dancing was, as may be imagined, very limited The festivities commenced with a true native dance, the performers being four men, with an equal number of women; the band, a rusty old fiddle played by an Esquimaux. The dance itself appeared to be a compound of a quadrille and double shuffle, interspersed with a few figures somewhat similar to those practised by the South Sea Islanders.

The seamen from the ships, and even some of our own party, eventually took a prominent part in the evening's entertainment, acquitting themselves with an amount of energy and grace, (?) accoutred as they were in heavy sea boots, that I was little prepared to witness. What with the closeness of the apartment and the perfume exhaled by the dancers, we were not sorry to leave the ball-room and reach the open air, though the proceedings did not terminate until a very late, or very early hour.

The settlement of Godhavn (Good Harbour), or, as it is generally called, Lievely, is one of those numerous but small colonies established by the Danish Government along the west coast of Greenland, whence they obtain a large supply of oil and skins, a trade over which they enjoy a strict monopoly. Each of these settlements is presided over by a Governor, or Chief Trader. These officers are Danes, and are immediately under the rule of the Inspector, who here reigns supreme, and whose word is law. The west coast is divided into two divisions, north and south, each of which is under the superintendence of an Inspector. Lievely comes under the charge of the Inspector of the Northern Division, who so far honours the place as to make it his residence.

The actual settlement itself consists of some half-dozen wooden houses, a church, and a few native dwellings of primitive construction, which might be more appropriately termed hovels.

The two principal houses are those belonging to the Inspector and Governor. They are neat-looking edifices, their interiors being clean and comfortable, denoting in each instance the presence of the fair sex. Next in importance to the Governor is the schoolmaster, who is about five-sixths Dane and one-sixth Esquimaux. In addition to his scholastic duties, he also officiates on Sundays in the little church, except on the occasion of the annual visit of the priest, who resides in another settlement, but who remains with them for five or six weeks every year. The church is a quaint little wooden building, having a small spire, and rejoicing in the possession of a bell! It contains about sixty sittings, has a small altar at its west end, and a pulpit in a corner at the side, giving one the idea that it had been placed in that position to be out of the way. I noticed a small concertina, to the strains of which the singing is conducted. Having no stove or fire-place of any description, it is not much frequented as a place of worship during the winter Divine service during those cold months is held in the school-house, which is about half the size of the church, and is heated by means of a stove. The remaining buildings comprise the smithy, the cooperage, and the brewery, besides a long stone store-house on the opposite side of the harbour, containing the provisions and stores sent out by the American Government for the use of the "Polaris."

The dwelling-places of the natives are most pleasing when viewed from the outside, and the greater the distance off the better. They are built chiefly of stone and turf, the only light during the summer months being that admitted through windows composed of the serous membrane of the intestines of the seal.

If sufficiently brave to encounter the offensive stench which pervades everything as to risk a visit to the interior, one passes through a long narrow entrance, having almost to crawl through upon hands and feet, emerging into a small room, not unlike the cabins on board very small and ill-found merchant ships, in which is the stove, the everlasting lamp, and the long bench or shelf on which they sit during the day, and on which, wrapped up in their skins, they sleep during the night. The number of people residing in one of these houses may probably amount to twenty or thirty of both sexes and all ages.

During the winter, when every aperture is carefully closed, and the fire and lamps kept burning day and night, the state of an interior may be better imagined than described, as the Esquimaux are notorious for being particularly dirty and filthy in their habits; in addition to which, the rotten and stinking pieces of seal and other animals that are left strewed about must largely contribute to the offensive stench that pervades their habitations.

The Esquimaux are a strong sturdy race, closely resembling in appearance the natives of Northern China. They have the same high cheek-bones and oblique-shaped eyes as the inhabitants of the Celestial Empire, but with thicker lips, and more full in the face.

They seem a good-tempered merry set of people, though decidedly deficient in the virtue of gratitude; taking things as a matter of course that may be given to them, and asking for things, no matter of what value, that take their fancy. The idea of giving anything in exchange is, with them, quite a visionary one.

This was strongly exemplified on board the "Arctic," as the good nature of our worthy captain made him load all that came off to the ship with cheese, pork, bread, soap (I do not think the latter, though readily accepted, is in great requisition), and various other articles, for which he received no thanks whatever. Indeed, some went so far as to complain of others receiving more than themselves, and requesting therefore a further supply. The women are by no means comely the prettiest part being certainly their costume, which is most picturesque. Their mode of doing the hair, which is tied up in a knot on the top of the head, is not prepossessing. They are very fond of any ornaments or outward show; beads and cheap jewellery are therefore in great demand. They seem very industrious, especially during our stay; when their time was fully occupied in making slippers, pouches, &c., from the skin of the seal.

It is most interesting to observe the skill and dexterity with which the men manage their *kayaks*, as their swift and fairy-like canoes are called, and the expert manner with which they use the harpoon. I saw one transfix a loom (guillemot) after a short chase, at the distance of twenty feet.

Disco, at the south-west extremity of which is situated the settlement of Godhavn, is a large island separated from the mainland of Greenland by a narrow passage, from three to four miles in breadth, called the Waygat. The traditions of the Esquimaux go so far as to say that it had originally been joined to the mainland, but that many centuries ago it had been broken off and towed to its present situation by a potent and influential *angekok,* or priest. They even go so far as to assert that the hole to which the tow-rope was fastened is the present harbour of Lievely. It certainly has the appearance of a round basin, being completely land-locked. Should their legends be true, it must have required a very powerful priest to transport such a large piece of land the distance indicated, its original site, according to the tradition, being close to Baal's river, about three hundred miles to the southward.

The hills and cliffs of Disco, bold, rugged, and precipitous, are composed of metamorphic rock, and rise in some places to an altitude of over three thousand feet above the level of the sea. The scenery is grand and majestic, and the effect is enhanced by the strong contrast afforded between the black rocks and their snowy mantle. In consequence of the great height of the land, rising abruptly, as it were, out of the sea, its distance is most deceptive when approaching from seaward. This deceptive appearance in these regions is frequently noticed by navigators, and fully accounts for the report brought to Norway by an old Norwegian skipper, or "famous sea-cock," as he is styled by the old chroniclers three hundred years ago, who, after undergoing many perils and hardships on his passage across to the east coast of Greenland, after sighting the land, seemed unable to approach, though the ship was steering directly towards it, and making fair progress through the water. At last, relinquishing all further attempts as useless, this "famous sea-cock" bore up and returned to Norway, reporting that his ship, after sighting the land, had been unable to advance in consequence of some hidden loadstone, which effectually barred his onward course.

Tuesday, June 3rd.—It was the captain's intention to have sailed this morning, but it had to be relinquished in consequence of a heavy snow-storm, the weather being too thick to proceed with any degree of safety. At 1 P.M., as it was still snowing hard, and therefore no prospect of our departure, I induced the doctor to accompany me on shore for a walk, my object being to proceed a little distance from the settlement and then ascend one of the hills in its vicinity. Arming myself with a long boat-hook staff, to serve the purpose of an alpen-stock, and the doctor taking with him his gun, we left the ship. We called at the Governor's house on landing, to leave some letters, and he attempted to dissuade us from our purpose, or, if we were resolved upon going, at least to take a guide; but being unable to procure the services of one, he kindly offered to accompany us as far as the mouth of the Red River, about a mile from the settlement, an offer which we gladly accepted.

Arriving at this place, he told us if we kept to the banks of the river we should be perfectly safe, as we could always retrace our footsteps, and bidding us good-bye turned homewards.

We continued for some little way skirting the edge of this small stream, though gradually ascending, until we had attained the estimated height of seven or eight hundred feet, when our attention was attracted by the rushing of water, and we observed away to our right a cascade descending over the rocks, evidently the source of the river along whose banks we had been toiling, whilst a deep and sombre-looking ravine branched away to the left.

As it was impossible to cross this ravine, the sides of which were not only precipitous, but in many places overhanging, we followed its direction to the left, continuing the ascent, which had now become rather laborious on account of the extreme steepness and the depth of the snow.

It was also necessary to use great caution, as we would frequently emerge on the very brink of the chasm, the snow being so dense as to render it impossible to see more than ten or twenty feet ahead. Had we only enjoyed clear weather, the scenery around us must indeed have been grand; as it was, the black and to us unfathomable yawning abyss, on the edge of which we would often unconsciously find ourselves, presented a sublime and terrible grandeur, such as I have rarely witnessed.

My companion, with more sense than myself, pointed out the insane folly of proceeding any further, observing, and very rightly, that in such thick weather there was no inducement to proceed, and that in all probability we should be unable to find our way back. I attempted to comfort him by saying that I had brought my pocket compass with me, and that we were therefore perfectly safe on that score, withholding the fact that we had taken a circuitous route, and that our compass bearings were, therefore, of no avail.

By half-past four, we felt from the force of the wind, that we had reached the summit of a hill, whose height we estimated, though of course very roughly, at over two thousand feet.

Deeming it imprudent, on account of the inclement state of the weather, to proceed any further, we turned to retrace our steps, but the snow was so thick that in a little time we could not see our tracks, and eventually lost ourselves. Affairs began to look serious as far as getting back that night was concerned; though we comforted ourselves by the knowledge that we had wherewithal to sustain nature until the snow ceased, the doctor having taken the precaution of putting a couple of biscuits in his pocket, and each of us being provided with a small flask containing brandy and water.

Knowing that we had come with the wind at our backs, and keeping close together for better security, we shaped a course head to wind.

The snow beating into our faces was positively blinding, making our eyes sorely ache. As it fell it froze upon our faces and clothes, and we were soon a complete mass of ice, though the exercise of walking, and the exertion constantly necessary to prevent ourselves from tumbling, kept us in a perfect glow of heat.

On one occasion I gave myself up for lost. Everything around was perfectly white, and it was impossible to say whether one was walking on a level piece of ground or on a steep decline. Suddenly my feet slipped, and in an instant I was sliding down the frozen surface of what I imagined to be the side of a glacier, which I was convinced would take me to the edge of the ravine and precipitate me into the gaping gulf beneath. I felt myself gradually gaining a greater velocity as I descended, when, providentially coming to a soft snow drift, I succeeded in driving my alpenstock deep into the snow, thus effectually stopping my rapid and headlong career. Looking round, to my horror I perceived the doctor directly in my wake, coming down at great speed; if he touched me I felt certain that we must both go, and I shuddered to think of the fate awaiting us. He was, however, like myself, enabled to dig his heels into the snow drift and was thus stopped. Putting the best face on the matter, we laughed heartily at our misadventure, and pushed on again head to wind.

Our mishaps were endless. Getting on to a nice firm piece, on which I was in hopes we should make good progress, I hailed my companion to follow, observing that it was "a famous bit for walking on." Hardly were the words out of my mouth before I plunged over head and ears into a drift, the doctor tumbling in after me. Extricating ourselves as best we could, we continued the descent, but for upwards of an hour we were completely lost, until by great good luck we suddenly emerged on the brink of the chasm along which we had ascended; and then we knew our troubles were at an end, as from thence we could with ease retrace our steps, ultimately reaching the settlement at about six o'clock; and hastening on board we were soon enjoying the luxury of a cup of hot tea and a change of clothing.

The weather having cleared a little, the captain determined upon proceeding to sea, and having bade farewell to our kind friends at Lievely, we were shortly after eight o'clock steaming out of the harbour, bound, to my great satisfaction, for the north.

During the afternoon, whilst walking along the beach on our return journey, we heard what sounded to us like the report of artillery. This we attributed to the rending and breaking up of icebergs, many of which were aground within a quarter of a mile of the shore. The sound closely resembles that produced by the discharge of heavy ordnance.

Esquimaux Dog.

Chapter VII.

UPERNIVIK AND MELVILLE BAY.

THURSDAY *June 5th.*—Snow has fallen without intermission during the last forty-eight hours; but this does not in any way impede our progress. Yesterday, at noon, we passed Omenak fiord, a famous place for icebergs, which, having been discharged from the glaciers scattered on both sides, accumulate at its entrance. The Danish settlement of Omenak is some little distance up the fiord. After threading our way through intricate passages caused by these floating ice islands, on some of which, in consequence of the thick weather that prevailed, we were nearly coming to grief, only avoiding collision by extreme watchfulness and caution, we passed Swarte-huk; and black enough appeared the land during the few short glimpses that we obtained. It was most tantalizing our being unable to enjoy what must really be grand and interesting scenery; but a thick snow storm made this impossible. Early in the morning we passed several islets, and at about eight o'clock made fast with an ice anchor to the land ice off the island of Upernivik, one of a group named by Baffin, during his remarkable voyage in 1616 round the bay which bears his name, the Vrouw or Woman Islands. The "Erik" followed soon after, making fast to a berg aground off the settlement.

Several Esquimaux kept pace in their swiftly gliding *kayaks* as we steamed in; and it was surprising to witness the velocity with which these frail-looking barques were propelled by their skilful occupants, who, with their long shaggy and unkempt locks falling over their shoulders, and their general wild and excessively dirty appearance, were more like some amphibious animal than human beings. Immediately the ship was secured we went on shore for the purpose of paying our respects to the Governor and Chief Trader, Dr. Rudolph, by whom we were most cordially received and hospitably entertained. What appeared to cause the good doctor the greatest annoyance was our incapacity, though *some* of our party acquitted themselves tolerably

UPERNIVIK.

Page 98.

well, to keep perpetually drinking port wine, sherry, rum, brandy, beer, and absinthe, all of which he insisted upon our partaking in an indiscriminate manner; and had we remained on shore longer than we did, the consequences would undoubtedly have been disastrous. No European intelligence had been received at Upernivik for nearly twelve months, so that the finding of Livingstone, the death of Napoleon, and the abdication of the King of Spain were all news to Dr. Rudolph, and these, with all other events of recent occurrence that we could remember, were poured into his willing ears. When informed of the death of Napoleon, his first exclamation, to our surprise, was, "What will my blacksmith say?" This man, it appears, who is a Dane, was an ardent and enthusiastic admirer of the late ex-Emperor of the French, whom he regarded as the first man of the age. The doctor immediately sent for the blacksmith, a fine stalwart smoke-begrimed looking man, to whom, after the ceremony of shaking hands and drinking port wine (it was about 9 A.M.) with the whole party had been gone through, he imparted the news of the death of the man he all but worshipped. I never saw change come over a man so quickly; he stared, evidently doubting his own senses, and when the words were repeated, the poor man burst into tears and rushed out of the room. When first seen he was in excellent spirits.

We walked over the settlement and visited two or three native dwelling-places, which resemble in every particular those at Lievely. The dogs did not seem to be so large or so strong as those of Disco, but they have the same wolfish appearance peculiar to the dog of these regions.

I had a conversation with the Governor regarding the chance of some of these animals being required in the ensuing year for an exploring expedition to the north. He informed me that in all probability enough could be provided at this settlement, though their value of late years had considerably increased. Captain Hall, of the "Polaris," had purchased twenty of these dogs, on his way to Smith's Sound, at exorbitant prices; some of the animals could not be obtained, though the sum of one hundred dollars was offered for each dog.

At half-past 4 P.M., having taken leave of the Governor, whom we honoured with a salute from seven of our harpoon guns, we got under weigh, and, accompanied by the "Erik," shaped a course through the thickly-clustered group of islands which are everywhere dotted about in this neighbourhood. Just as we were losing sight of the settlement, we noticed a puff of white smoke, near the flag-staff from which the Danish ensign was displayed, then another and another, followed by a mild report, which informed us that our salute had been acknowledged and returned by three guns. It was more than probable that these guns, with their ammunition, had to be dug out from some little nook or corner in an out-of-the-way storehouse.

We followed in the wake of our consort, whose experienced commander, being well acquainted with the intricacies of the navigation through these islands, had volunteered to lead the way, until shortly after 7 P.M., when a thick fog overtaking us, compelled us to stop, each ship making fast with three ice-anchors to some land-ice in a bay of one of these islands.

We were within half-a-mile of the shore, whose precipitous cliffs rose up to a height of eight hundred or one thousand feet.

The ice on which the men had to step to get the anchors to the fixed ice beyond was of a very soft and treacherous nature, and many were the immersions that took place whilst carrying out the necessary duties of securing the ship. A dip in the water with the temperature two or three degrees below the freezing point is no joke; yet all laughed heartily when an unfortunate individual went through, and even the victim himself, putting the best face on the matter, seemed to enjoy the fun as much as any one.

In leaving Upernivik, we quite bid farewell to the outward world; it is the last place at which we have an opportunity of leaving letters for England, and, with the exception of occasionally meeting a whaler, it is the last place at which we shall see and converse with civilized beings.

Upernivik is the most northern settlement of any importance on the coast of Greenland, and is in fact the emporium of all the others in its vicinity. Tesuisok, Kingitok, and Susak, are the only ones to the northward, and these are all situated in the Woman group.

The Governor, Dr. Rudolph, who has resided in Greenland for thirty-five years, collects the produce of oil and skins from the different settlements, rendering an account of everything to the Inspector of the Northern Division.

He has under his immediate rule over six hundred people, though the population of Upernivik itself numbers only seventy-nine; out of which eight are Danes, including the Governor, priest, and blacksmith, and their wives and families, the remainder being Esquimaux. Like Lievely, it has its church, store-house, cooperage, smithy, &c.; but, unlike Lievely, has no good harbour in which a vessel could remain with any degree of safety. True, there are always bergs and land-ice to which a ship can be secured, but, on this occasion, it is necessary to keep the steam ready for any emergency.

The scenery about these islands is wild and peculiar, exhibiting alternate patches of bare gneiss, and snow scattered upwards to the summit of the hills, which rise to a height of from one thousand to fifteen hundred feet.

Friday, June 6th.—Weighed shortly after five this morning, and, still in company with the "Erik," proceeded towards the north; passing through large and extensive streams of loosely-packed ice, and threading our way amongst the numerous islands off this part of the coast.

Icebergs of large dimensions were in all directions, which added to the beauty and novelty of the scene, enhanced as it was by a glorious bright sunshiny and real arctic day.

At noon we passed the Horse's Head, (why so named I have been unable to discover, as it certainly bears no resemblance to one,) steaming within a quarter of a mile of Cape Shackleton, a precipitous cliff rising abruptly from the water to an altitude of fourteen hundred feet, a famous place for looms; indeed, this place has the reputation of possessing one of the largest and most prolific loomeries on the coast of Greenland.

The cliffs have the appearance of the same metamorphic formation as at Disco, although striated in a diagonal direction.

From Cape Shackleton we emerged into a large open water, entirely free from ice, which astonished all on board, promising a fair and easy passage through Melville Bay.

Taking advantage of this favourable opportunity, we steamed on, passing close to the Deer Islands, off which there is a remarkable and curious-shaped hill, called by the whalers "Kettle bottom-up hill." When viewed from the southward, it certainly bears some resemblance to that useful article of kitchen furniture; but when seen from the northward, it takes the form of a sugar-loaf, being high and truncated. Off the Duck Islands were several large bergs aground, between two of which, whose summits towered far above our mast-heads, we steered, beholding with admiration their wonderful structure. One was most beautifully adorned on the side nearest us with an ornamental fringe of icicles, pendant from its overhanging brow. Wilcox Head, which may be regarded as the south extreme of Melville Bay, was next sighted, and by 7 P.M. we were abreast and in sight of a singularly-shaped hill, called the "Devil's Thumb," appearing like an obelisk on the distant land.

Now our troubles are about to commence, for stretching out from the nearest point of the shore to the northward and westward, as far as the eye can reach, is our great enemy, the dreaded floe ice of Melville Bay.

Imagining we saw a "lead,"[1] closed only by a narrow neck of ice, we endeavoured by "butting" and "boring" to break through this obstacle; but though we tried for a couple of hours, we were unable to advance a single ship's length, and we were compelled to retrace our steps and search for a lead elsewhere.

To the northward of us, and therefore ahead, we can see the smoke and the top-gallant yards of another whaler, just visible above the icy horizon, but the floes have apparently closed up the passage by which she had gone through, so that the knowledge of her position is of very little value to us.

Most ominous and significant preparations were made during the day. Provisions were hoisted up from below, and ranged along the upper deck, in readiness to be placed in the boats, or thrown out on the ice, should it be necessary to abandon the vessel, each man of the ship's company being ordered to have a shift of clothing packed up

---

[1] A "lead," as it is termed, is an open stream of water between the floes, through which a vessel is able to pass.

handy in a small bag. When a vessel is "nipped" by the ice, there having been no time to cut a dock out of the land floe, this relentless foe must pass either *over* or *under* the ship, and that so quickly that the men have barely time in the former case to jump out on the ice. Several instances have occurred in which the ice has nipped and gone *through* a ship; and on one occasion, that of the whaler "North Britain," in the year 1830, it is related that the surgeon, who was sitting in the cabin at the time, beheld the ice breaking through *both* sides of the ship, and he was barely able to make his retreat in safety. In that year no less than twenty vessels were lost in Melville Bay, some of them being literally crushed to pieces. The year 1819 was also most disastrous to our whaling fleet, fourteen vessels having been lost; in fact, until the introduction of steam, scarcely a season passed without the destruction of many vessels. The water below the ice in Melville Bay could indeed unfold a sad tale. Many is the stout ship, manned by a daring crew and commanded by a skilful and brave master, that has perished, crushed into innumerable fragments by the insatiable and ponderous floes rapidly closing upon the unlucky and doomed vessel, swallowing her so rapidly as barely to allow time for the escape of the crew.

It is very remarkable that these frequent casualties are seldom or ever attended with loss of life. The real danger to be apprehended on that score is from insufficient food and exposure in a rigorous climate. After the mishap has taken place, if there are no vessels in company, the shipwrecked crew must needs find their way to the nearest Danish settlements, and this can only be done by means of boats or sledges; in the latter case only when the ice is of sufficient thickness to render travelling on it perfectly safe.

In the year 1830, which, as has already been mentioned, was fatal to twenty of our whalers, two vessels, the "Princess of Wales" and the "Letitia," were destroyed by the ice passing completely through their broadsides and literally cutting them longitudinally into two. The "Resolution," of Peterhead, had the whole of her counter pierced, and eventually sank; the "Laurel" and the "Hope" were squeezed perfectly *flat*, and then thrown violently on their broadsides, and the "Commerce" (brig) was lifted bodily on the ice with a twisted stern post, and sank directly the pack loosened. The "Baffin," "Achilles," "Ville de Dieppe," and "Rattler," were crushed to pieces, and the "Progress," of Hull, was totally destroyed by an iceberg. The other vessels were destroyed in various strange ways.

Since steam has been introduced into the whaling trade, the dangers of Melville Bay have been much reduced, and vessels have of late years been enabled to navigate those waters in comparative safety. I make use of the expression "comparative safety" advisedly, for of course there are instances in which even steam vessels have been lost; but these are happily very rare.

Some seasons are more severe than others, and again some are more open. Cases occur in which vessels, though possessing the power and advantages of steam, have been unable to effect a passage through the bay, and have been compelled to return. The little "Fox," in 1858, under the command of that most experienced of Arctic navigators, Sir Leopold McClintock, was not able to reach the North Water,[1] and was forced to pass a dreary winter in the pack; and only as late as last year a large and powerful steam whaler, though commanded by one of our best ice navigators, was unsuccessful in her endeavours to penetrate the ice in Melville Bay, and had reluctantly to retrace her steps and seek for whales in Davis' Straits.

The first ship that ever attempted to brave the perils and difficulties of this dreaded passage was

[1] By the "North Water" is meant the open water into which the vessels emerge after passing through the ice in Melville Bay.

the "Larkins," of Leith, in the year 1817, followed by the "Elizabeth," of Aberdeen; since which time, with very rare exceptions, the North Water has been reached by the greater part of the vessels, and generally by the whole of those comprising the whaling fleet. In the early days of the whale fishery, and before our vessels were so well adapted to encounter the hard and rough usage inseparable from ice navigation as they are at the present day, that is between the dates 1817 and 1849, there were only four years in which attempts to reach the North Water were totally unsuccessful.

The latest period at which whalers have persevered in their attempts to make a passage through Melville Bay, terminating with a successful issue, was early in August.

The time occupied in reaching the North Water must necessarily vary according to the season; for we find that Parry, in 1824, was fifty-four days passing through Melville Bay, whilst Sir John Ross, in 1829, was only five days! Captain Austin's expedition, in 1850, was forty-five days making the passage, whilst, two years after, Captain Inglefield, in the little "Isabel," went through without any detention whatever.[1] Now, our

[1] Since writing the above, I have learnt that the "Polaris,"

whaling captains consider themselves very unfortunate when they are detained for a period of fifteen or eighteen days.

Saturday, June 7th.—To-day finds us fairly battling with the ice. We made excellent progress during the night by keeping along the southern edge of the land floe, taking advantage of the different "leads" or lanes of water as they presented themselves, by which we were enabled to advance some distance. The "Erik" was still in company, each ship taking the lead, according to her success in finding a passage or otherwise. At 7 A.M. we passed the little "Victor," the vessel we had observed ahead the previous night, struggling bravely on; but, not being possessed of such power as ourselves, she was soon left astern.

The wind is blowing fresh from the southward: this is from the very worst direction that we can have it for crossing the bay, as it packs the loose ice tight up against the land floe. It is what the whalers call a "strong ale wind," as, during the time it blows from that quarter, ships are more liable to get "nipped," and therefore destroyed.

---

commanded by the late Captain Hall, also passed through Melville Bay without the slightest obstruction. In fact, no ice of any consequence was met with until they had attained their highest latitude, namely, 82° 16 N.

The reason it is so called is, that directly the captain has announced his intention of abandoning the ship, the crew rush aft, provided there is time, possessing themselves of anything that may please their fancy, and, sailor-like, immediately broach the ale and spirit casks, unless the captain has wisely beforehand spilled the liquor by staving the casks. This eager thirst for strong drink has often perilled the life of a good and brave seaman.

Several times during the day did we push up a likely looking "lead," seeing a splendid and encouraging water sky[1] in the distance; but as often had we to turn back, and quickly get clear of the ice to avoid a "nip," the floes being in such rapid motion.

From the nest a stream of water would be seen, which in less than fifteen minutes, a less time than we could possibly steam through it in, would be completely and hopelessly closed. To add to our difficulties, the weather became very thick, snow falling heavily, and this necessitated an immediate halt, so the ship was made fast at about 1 P.M. with a couple of ice anchors to a large floe; the

---

[1] By a "water sky" is meant a peculiar bluish colour in the sky, which always denotes the presence of water. On some occasions it is seen over the ice at a distance of many miles, and is a sure sign of open water.

So also the "ice blink," or a whitish colour in the sky, is a certain sign of ice.

"Erik" following our example and bringing up about two cables length from us. At about 6 P.M. our little squadron was augmented by the arrival of the "Victor," which made fast to the eastward of us.

In the evening we received a visit from her captain, and I got some little insight into the mysteries of a "mollie," though on a small scale.

In whaling parlance, a "mollie" means having a night of it; that is, a number of captains congregate together on board one ship, and then an animated discussion ensues regarding the success attending each and every individual engaged in the fishery; as talking is naturally a thirsty occupation, copious libations of spirits and beer are discussed, forming very important items in these orgies, which on several occasions have terminated in anything but a friendly manner.

During the fall fishing, that is towards the close of the year, I have been informed that a "mollie" will last many days! It derives its name from the fulmar petrel, a bird, as I have already stated, that assembles in great quantities during the operation of flinching a whale, when they are continually fighting, squabbling, and gorging to an inordinate degree, over the delicate morsels that are carried away on the water from the ship or fish during that process.

Tracks of bears were seen on the floe during the afternoon, but Bruin very wisely kept himself scarce.

Sunday, June 8th, was ushered in by a bright sun, of which I took advantage to land on the floe with my artificial horizon, and determine the position of the ship by double and single altitudes. I find Captain George's artificial horizon[1] a very useful and portable instrument, and well adapted for these regions, in consequence of the very simple method by which the mercury is transferred from the bottle to the trough and *vice versâ*: no easy task with the common roof horizon, when the temperature is several degrees below the freezing point. To-day the thermometer was as low as 25°

Shortly before noon, the captain seeing what he considered a good lead, though separated from us by a broad stream of ice, determined upon boring his way through. Steam was raised, the ice anchors taken on board, and at the apparently unyielding ice we went full speed. Men were stationed on the floe, on each bow of the ship, whose duties were to remove with long boat hooks and handspikes the fragments of ice as they were

---

[1] Supplied by Messrs. Gould and Porter (successors to Cary), of 181, Strand. All the instruments supplied from this firm I found very good.

broken off and crushed by the force of our blows. The ice in many places was several feet in thickness, and the floes had long projecting tongues under water, commonly called *calves*, and these catching the fore foot of the ship would frequently turn her head round, and consequently away from the passage aimed at.

Pieces of ice of considerable magnitude, broken off by our repeated charges, would pass along under the bottom of the ship, and, emerging up the screw aperture, prevent the propeller from performing its revolutions, and therefore bringing the engines to a dead stop. Those of the ship's company remaining on board were all this time employed in running from side to side on the upper deck, for the purpose of rolling the ship and thus crushing the ice and making a free passage for us to pass through.

Our exertions were rewarded with success, and we soon had the satisfaction of emerging into a clear lead, through which we steamed into a fine open water. The other two vessels, observing our success, immediately took advantage of the opening we had made, and were soon in our wake.

Several large ground seals were seen lying on the ice in divers directions, and also, later in the afternoon, a couple of bears, who, however, evaded our futile attempts to make a capture. It is surprising to witness the wonderful speed and celerity at which these animals journey over the snow and ice, scampering away as if the surface was as smooth and level as a bowling-green.

At five in the afternoon, snow commenced falling heavily, obliging us again to make fast to a floe. Clearing up slightly at about half-past six, our skipper, who is determined not to allow the grass to grow under his feet, weighed and stood away along the edge of the floe on the look-out for a "lead."

At eight, after passing through an immense quantity of loose pack, we came out into a large open water, where we felt a considerable swell, so much so that the boats at the davits had to be belted to and freshly secured. Can this be the North Water? Are we really clear of the whaler's bugbear, Melville Bay? The attainment of such a desirable object seems almost too good to be true.

The weather is still very thick, accompanied by a driving sleet and snow, with a fresh "strong ale wind." This, however, if we are clear of the bay, is no detriment to us. The "Erik" is nowhere in sight, and it is long since we have seen our little friend the "Victor." Both yesterday and to-day myriads of rotges or little auks were seen, the air and water in places literally teeming with them.

As we approached in the ship to where they were swimming and feeding, these little birds would instantly dive, and we could distinctly see them in the clear water swimming, or, as the sailors aver, flying, at the depth of three or four fathoms. The harsh discordant noise which these birds give out at times produces such a din as effectually to preclude the possibility of hearing any other sound.

Several dovekies, or black guillemots, were also seen.

Little Auks.

CHAPTER VIII.

THE NORTH WATER.

MONDAY, *June 9th.*—A rattling southerly wind, before which the "Arctic" is going nine knots, snowing at intervals during the day, and the temperature as low as 23° We have made a wonderfully quick and almost unprecedented passage across Melville Bay. All last night and early this morning we kept along the outside edge of the land ice; and at about 3 A.M. we picked up the eastern edge of the middle pack-ice, with a fine clear open water between, though occasionally meeting and having to force our way through broad streams of straggling ice. At 5 A.M. Cape York was in sight, and by ten o'clock passing Conical Island, we were off Cape Dudley Digges, with Cape Athole and Wolstenholme Islands on our starboard bow.

We are now fairly in the North Water; but though we have been wonderfully successful in passing through Melville Bay with only a few hours' detention, I cannot but acknowledge to a slight feeling of disappointment in not having witnessed a few of those dangers and difficulties experienced by many whilst performing this dreaded passage. I was also most anxious to see the "crimson snow" on the hills of Beverley; to land at Cape York and communicate with those interesting Esquimaux, styled by Sir John Ross the Arctic Highlanders; to shoot deer in the neighbourhood of Cape Athole, where they abound; and to bag some eider duck and obtain their eggs amongst the islands where they congregate in thousands. The glaciers also, which were distinctly visible from the ship, extending far away into the interior, were objects of great interest to me, and of which I was most anxious to obtain a closer inspection.

At about 2 P.M., being in latitude 76° 20′ N., with the Cary Islands in sight on our starboard bow, we reached the northern limit of the middle ice, and our course was altered to the S.W., our fishing ground being off the entrance to Lancaster Sound.

Only 850 miles from the North Pole! It seems

no distance. To the northward appears a fine open water, interrupted only by a few insignificant streams of straggling ice, extending out from Whale Sound.

One day's steaming would take us to the portals of Smith's Sound, that mysterious region by which the vast extent of unknown land around the Pole seems alone attainable.

What a wonderful man was old Baffin, who in his small and crazy vessel of 55 tons so successfully penetrated these then undiscovered regions! He must indeed have been a brave old navigator, and one whose deeds we should strive to emulate.

I am quite convinced, and the captain is of the same opinion as myself, that *this* would have been a splendid year for discovery. We have had four remarkably open seasons in succession, and with such a ship as the one I am now in, there is no saying what such men as McClintock, Richards, or Osborn would not perform. It almost seems an opportunity thrown away that may possibly not occur again for some time, though we must remember that should *this* summer be followed by a mild winter, our prospects of reaching a high latitude *next* year will be much increased. If the entrance to Smith's Sound appears so free from obstruction as it does now, what will it not be in two months hence?

We make short work of our degrees of longitude, having rattled off ten during the last twenty-four hours.

Tuesday, June 10th.—We still continue to make fair progress, though compelled at times to force our way through broad streams of straggling ice, and through many miles of pancake or bay ice, which is young ice formed by the heavy fall of snow and low temperature that we yesterday experienced. At 2 A.M. we came up to where the water was swarming with innumerable rotges: a couple of boats were lowered, and four guns sent away to shoot for the "pot." They returned in less than twenty minutes with between three and four hundred. The little birds were swimming and flying in such thick clusters that *forty-five* were killed at one discharge from a gun, and *thirty-three* at another! It sounds very much like murder.

The flesh of these birds is excessively sweet, and they afford a very pleasing contrast at our meals to the continual beef and potatoes. At 11 A.M. we caught a glimpse of Cobourg Island, situated at the entrance to Jones' Sound, and by 3 P.M. were off Cape Horsburgh.

Occasionally, during intervals between the snow

showers, we succeeded in getting a peep at North Devon, so called, I presume, from its extreme dissimilarity to the coast along the north side of our own beautiful county, for it was apparently the very picture of barrenness and desolation, and covered with snow. Everywhere snow! Still skirting along the edge of the middle ice, we arrived at what was considered the northern limit of our fishing ground, off the entrance to Lancaster Sound. Fires were burned down, and the ship, for the first time since leaving the S.W fishing, once more put under canvas.

August is generally regarded as a blank month, so far as regards the whales, by those employed in their capture, very few being seen during that month. The captain and myself have repeatedly had discussions on this subject, in which Baffin, Inglefield, and Hayes have been quoted. All these authors testify to the fact of having seen numerous whales in Whale Sound in August. It was, therefore, with difficulty I could repress my delight when the captain came up to me to-day, and announced his intention, if he is not a full ship, of going after "them beggaring whales" in August, which, in his opinion, go north, and therefore he would search for them in Whale Sound. This conjecture was of course readily assented to by me.

It is very strange that the masters of whale ships, with very few exceptions, still continue to fish over the same old beaten ground, never attempting to depart from the old routine, and, therefore, allowing the month of August to pass idly, without troubling themselves to discover the resort of the whales during that month. I am firmly convinced that our captain s enterprising disposition will shortly resolve that question, and, whether we go north or whether we go west, we shall eventually return to Scotland a "full ship;" and I am confidently looking forward to a prosperous and successful cruise, bringing with it important and useful results.

This evening we had for tea a most savoury mess, consisting of some preserved ptarmigan, which had been kindly presented to us at Lievely by Mrs. Smith, the Inspector's wife.

Wednesday, June 11th.—When Jack, that most obliging and willing of stewards, called me this morning he imparted the not very pleasing information, that it was "blowing a gale of wind, snowing hard, and freezing like mad." With such intelligence it required no small amount of submission and fortitude to abandon my snug warm little bunk and perform the necessary morning ablutions. My toilet completed and chronometer wound up, I proceeded on deck, and found things not so bad as I anticipated from the steward's report. It was blowing fresh from the S. E., and the ship was carrying reefed topsails and foresail, though the water, on account of the ice, which was scattered about in loose straggling streams and detached floes, was as smooth as the Thames off Westminster on a boisterous day. Snow and sleet were falling fast, but the temperature was only 6° below the freezing point. On the whole, perhaps, the day had not a very inviting or promising appearance.

We are fairly on our fishing ground, and every one is on the *qui vive*: boats are lowered square with the gunwale, oars counted, thole pins and grummets inspected, and the harpoons and lances, on which a thick coating of rust has lately been allowed to accumulate, cleaned and sharpened.

Midnight.—I am now writing amidst a perfect chaos of empty bottles, broken pipes and glasses, and cigar ashes, inhaling a strong perfume of stale tobacco. And the cause of all this is, we have had a "mollie." It has now terminated, and I am left in sole possession of the cabin, attempting to write on the table, garnished in the manner described.

But, to account for our dissipation, I must state,

that, at about 1 P.M. the weather, with one of those rapid changes so peculiar to these regions, suddenly cleared up, revealing to our view the land along the south side of Lancaster Sound, high and covered with snow, and around us in different directions no less than five whalers.

A boat was lowered, and we went on board the "Esquimaux," which ship we knew had left Dundee a fortnight after us, receiving from her kind commander a large bundle of letters and newspapers. Making our visit as short as was compatible with courtesy, we hurried on board, and were soon deeply engrossed with news from home.

At about 7 P.M. our hospitable skipper hoisted a bucket at the mizen-top gallant mast-head, which, it seems, is a signal, when no boats are away, that the master of the ship displaying such bucket is desirous of having a "mollie." In a very short time we were boarded by the various captains, some of whom were accompanied by their surgeons. Then ensued the usual scene, plenty of talk about whales; plenty to drink and plenty of smoking. From the masters downwards all seemed to be known, and their individual merits and qualifications were severally discussed.

The topic of conversation was of course the whale fishery; no fish has as yet been seen, but they all seem to concur in one opinion, namely, that it is a fine open season, and they are unanimous in predicting a successful slaughter amongst the whales. They had all a good passage through Melville Bay, and are only two or three days ahead of us. From conversation that I had with some of these captains I learnt with great satisfaction that they already deeply regretted having signed, some few months previous to their sailing, a paper at Dundee, advocating the route by Spitzbergen as being the best adapted for reaching the North Pole. They were drawn into the act by one of the whaling masters. They are now entirely in favour of the route *viâ* Smith s Sound, which they are most decidedly of opinion, and they are willing to give their ideas publicity, is the best and, perhaps, the only way by which the Pole may be reached.

Saturday, June 14th.—The day before yesterday, seeing a large open water some distance to the southward, steam was raised, and leaving the remainder of the whaling fleet, we forced our way through the ice, eventually entering what appeared a splendid "water."

The floes, however, were fast closing, and before we had time to extricate ourselves we were completely and for the time hopelessly beset. We used our utmost endeavours to get clear, sail was set, and the engines did their best; but it was of no avail, we were unable to bore a way through, and there we were, regularly caught and unable either to go ahead or astern. The ice was so soft and brashy as to render it impossible to benefit ourselves by cutting or blasting.

In addition to the mortification that we feel at being beset, and probably detained many days, we have the pleasing knowledge that the whales are beginning to make their appearance, and that the "Camperdown" succeeded in getting a fish at two o'clock yesterday morning, the first that has been captured since leaving the S.W

Yesterday was a beautiful clear day, and our view from the ship across a large and extensive plain of ice to the thickly snow-covered hills beyond, rising to an altitude of fully 2,000 feet, was both new and imposing. Not a patch approaching a dark colour was to be seen to relieve the eye on the pure white snow-clad mountains.

Close to Cape Liverpool is a large glacier, which we can plainly see, formed in a deep ravine between two high hills, which makes me anxious for a trip to the shore. There are also other inducements to tempt me shorewards. Natives might be fallen in with who could possibly throw some light on the unknown regions of the north, or from whom some information may be obtained regarding the missing "Polaris," the Esquimaux, as a general rule, being of a most migratory habit. Positions can be determined, and specimens in various branches of science can be collected. Altogether there are many attractions to draw one towards this little-known land, to obtain any one of which would amply repay one after the monotony of life on board.

This morning brought a very imperceptible alteration in our prospects, as far as the chance of getting clear was concerned. Snow was falling heavily, and the wind had subsided. At about noon the snow ceased, and a bright sky and warm sun took its place, but not a drop of water to be seen in any direction.

The land about Cape Byam Martin extended along our starboard beam, but everywhere else was ice—one impassable plain of ice. We made several efforts to release ourselves, but after nearly two hours' steaming these were relinquished, having only succeeded in that time in moving the ship about half her length.

The ice is of that brashy nature which precludes all possibility of walking; and being thickly covered with snow, it would be excessively dangerous for any person to venture upon it. Anyone so doing would be almost certain of going through.

This fact is apparently known to an old seal of

a huge size, who for many hours has been lying on the ice basking in the sun, occasionally raising its head for the purpose of observing our movements, in a dreamy, sleepy sort of manner. We could easily shoot him from the ship; but even if successful in killing him, it would be almost impossible to make a prize of him, so he is allowed to remain in undisturbed peace, and enjoy himself, as Dr. Kane would say, in a "sealy" manner.

These regions are proverbially notorious for the sudden changes which so quickly take place in the weather and state of the ice.

At noon our position was by no means a pleasant one, but by 4 P.M., in consequence of a bright sun and a light north-westerly breeze, the floes commenced to loosen and "leads" through the ice were observed in various directions. Steam was quickly raised, and hopeful anticipations of being again free were shared by all. These were, however, doomed not to be realized, for after about six hours' continual steaming, first ahead, then astern, with the men employed rushing from side to side to roll the ship, we were again beset, having barely advanced a couple of miles. If the fine weather continues we may reasonably hope to be shortly released; and the sooner the better, for our confinement and utter inability to get out are already beginning to exert a depressing influence upon our worthy skipper, who is constantly picturing to himself the other ships in the midst of whales, whilst we are lying idle; and he is, therefore, rather morose and taciturn, so different from his usual jovial and boisterous humour.

Bear sleeping.

CHAPTER IX.

MIDDLE ICE FISHING.

SUNDAY, *June* 15*th*.—Last night, before going to bed, an inquisitive little seal made its appearance in a small pool of water astern, seeming to be monstrously surprised at beholding such a clumsy and awkward-looking animal as we must have appeared in comparison to itself. Poor little fellow! it paid dearly for its temerity. It is what the whalers call a "floe rat," is of a dark-greyish colour, and is about three feet long.

Our days, whilst fishing, are most uncertain, and to-day has been no exception to the general rule. What would our strict Sabbatarians in Scotland say, if they only knew how their countrymen had been employed during the last twenty hours?

Early this morning and during the entire forenoon, taking advantage of the looseness of the pack, we pushed our way through the cracks and lanes, and by dint of great patience and perseverance, succeeded by noon in extricating ourselves from our icy prison, emerging into a fine piece of water, not far from two other whalers.

At about half-past two a fish was seen from the masthead, and shortly after another. Six boats were promptly in the water and in pursuit, when more fish were seen, and all the boats were despatched in chase. In a very short time the welcome cry of "A fall!" was called, and we heard it re-echoing along the water from the various boats spread out in different directions.

Again the cry, and then another and yet another, and before five o'clock our boats were fast to no less than four fish! Our success must have been most tantalizing and vexatious to the two ships in company, who, though amongst the whales with the same opportunities as ourselves, failed to secure a single fish. Our boats were scattered in all directions, and fog and snow overtaking us, they were soon concealed from our view. We were, however, guided to a knowledge of their position by hearing the cheers of the victorious crews when they had succeeded in killing their fish.

One of the whales, on being struck, had made for the pack ice, which rendered it impossible for a boat to approach sufficiently near to administer the *coup de grâce.* Steam being ready, the ship was pushed in, steering towards the spot where, from the direction the line was taking, we knew the fish to be, making thereby a lane for the boats. We came close up to the poor beast when a second harpoon was fixed into it, the harpooneers watching their opportunity every time it rose, and plunging their lances in to the depth of six or seven feet. The water was soon dyed crimson in the vicinity of the unfortunate animal, and the ice for some distance around was stained with the same ensanguined hue. It was impossible for anything possessing vitality to survive long the vigorous and deadly thrusts which it received, and after a brief struggle and one last convulsive heaving of its enormous tail, the unhappy monster, yielding to the superior power and knowledge of its tormentors, turned slowly over on its back and expired amidst the cheers of the boats' crews and all on board. It was impossible to refrain from joining in the general excitement and enthusiasm of the moment, though painful to witness the death struggles of these huge, unwieldy animals.

It was eight o'clock before all our fish were killed and alongside, when everything being prepared for the process of "flinching," the men were sent to their supper, and at nine o'clock commenced that operation.

The fish at this early period of the season are generally small, which can only be accounted for by the supposition that the young fish always precede the older ones on their journey up from the south. In consequence of their small size, the operation of "flinching" is slightly different from that which I have already described. No "cant" is used on these occasions, but the fish is taken in, in four hoists, with the fore and main spek tackles. This process is called "worming." The first hoist is the "jowl," which is the entire under jaw, including the tongue ; the second is the "crown," which comprises the upper jaw and the remainder of the head ; this will of necessity include the whalebone ;—the head of a whale is about a third of the whole animal. The third hoist is one fin and all the blubber from the belly and one side; and the last hoist is the remaining fin, tail and blubber from the back and other side, the carcase sinking as the last piece is detached from it. It was past midnight before our work was finished, and the last *kreng* disappeared below the surface of the water, after which we all sat down to a supernumerary meal, consisting of hot preserved Australian mutton and coffee, which of itself, exclusive of the bright daylight which now prevails, would effectually be the means of banishing sleep.

The fulmars and burgomasters, as usual during

the process of flinching, held their noisome and disgusting orgies, gorging themselves to such a state of repletion on the refuse pieces of *kreng* and blubber, as in several instances to be unable to fly away.

One of the fish captured was completely blind of one eye, the eye-ball being simply a mass of fat and gristle. It had not the appearance of being diseased, and in all probability had never been a perfect organ. The ear of the whale, or rather the aperture leading to the tympanum, which is situated just behind the eye and about two feet inside the head, is remarkably small, not sufficiently large to admit the insertion of a man's little finger. The drum of the ear itself, when extracted and cleaned, bears a strong resemblance to a large univalve shell.

Whilst hoisting in one large mass of blubber, the strap to which the tackle was hooked, in consequence of the heavy strain, drew through the soft, pinguid substance, which latter falling backwards into the "mollie boat," knocked poor Tom Webster overboard into an unctuous sea of water, blood, and grease, out of which he was picked in a most deplorable plight, retaining, however, his pipe in his mouth, though losing a valuable spade with which he had been working. A bath with the temperature below freezing, whilst perspiring from over-exertion, is no joke.

It is surprising to notice the exhilarating cheerfulness which appears to take possession of everyone on board after a successful bout with the whales, and I must also myself plead guilty to the fact of sharing in the same exultant feeling.

Monday, June 16th.—"After a storm comes a calm," a saying which has been verified by the perfect quiet of to-day, the general excitement of yesterday having totally subsided.

Snow was falling the greater part of the day, and the weather was thick and foggy. During the forenoon a large walrus, or sea-horse, as they are more commonly called, was seen swimming about at some little distance from the ship. I fired at it with my rifle, but only succeeded in wounding it. Indeed, it is almost impossible to kill these animals with a bullet, their skulls, the only vital part, being of such a thickness as to make them in many cases impervious to a common rifle ball. I was sorry after I had fired, as the poor beast was seen for some time splashing and writhing with pain, though his sufferings will most probably terminate with death before many hours elapse.

A slight commotion was caused towards the evening by observing the boats of one of the whalers fast to a fish. Instantly smoke was seen in all directions, and like a flock of harpies, the whole of the whaling fleet, ourselves included, were steaming towards the lucky ship, determined either to share in her good fortune—supposing that she was in the midst of fish—or to scare them all away. Nothing, however, resulted from our efforts.

Tuesday, June 17th.—This whaling is a very selfish trade; if one ship is unsuccessful, nothing delights her master and crew more than to observe others equally unfortunate. This has not been one of our lucky days; and although we have seen several fish and had our boats away in pursuit, we have not been able to record a single "fall," whilst the other ships appear to have been as prosperous as we have been the reverse.

It is a lovely day, and the land stretching away from Cape Bathurst to Cape Liverpool is plainly visible about ten miles off, still retaining its thick winter coating of snow.

Taking advantage of the fine weather, the crew have been employed "making-off" the blubber from the fish captured on Sunday; and grease—good thick substantial grease—reigns supreme. The deck is as slippery as ice, and the unctuous substance pervades the whole ship; so much so, that great caution is necessary in walking along the upper deck to avoid a tumble. The smell is not so objectionable as I had anticipated—perhaps I am getting so accustomed to it as hardly to perceive it.

Apparently one of the most happy and contented of our crew is the "Dummy," as he is always called. Notwithstanding his infirmities, he is a most intelligent and willing man, besides being of a very cheerful disposition. He is only able to make himself understood by signs peculiar to himself, never having learnt to talk with his fingers, or to make use of any recognized deaf and dumb alphabet.

He always appears to know intuitively when boats are ordered away, or any other duties connected with the working of the ship, and is generally among the first The only noise he is able to emit is a disagreeable sort of croak, which sounds oddly enough when he is attempting to join the men in their incessant cheering on the death of a fish and other occasions.

Wednesday, June 18th.—How rapidly changes take place in these regions! Four days ago we were beset in much about the same place as we are at present, by closely packed ice; now the ice is of a straggling description, and the water, as far as the eye can see, perfectly navigable.

During the forenoon a couple of narwhals, or as they are called by the whalers, "unies" (unicorn abbreviated), were seen not far from the ship, and the captain gave permission for a crew of volunteers to take a boat and attempt the capture of one. I had the honour of being selected as boat steerer—no unimportant duty—the success of the harpooneer mainly depending upon the manner in which a boat is steered towards a fish, the general maxim being "to keep off her eye." It also requires a little skill and dexterity, gained only by constant practice, to manage a steer oar in an efficient manner.

Pulling cautiously up towards our quarry, we had to wait some little time before a chance presented itself; it came at length, and Harky[1] Hunter, our harpooneer, immediately profiting by it, fired,—the line running out rapidly and the water discoloured with blood, plainly told us with effect. It ran out 120 fathoms of line with great velocity and without a check, when it suddenly stopped. We commenced hauling in the line, coiling it away carefully as it came in, and soon had the satisfaction of seeing our victim come to the surface dead. It had probably been drowned in its death struggles. It proved to be a valuable prize, measuring eighteen feet in length, exclusive of the horn, which appendage was seven feet long. These creatures, when struck, will take line out as quickly as a whale, though not such a quantity.

On our return to the ship, the "unie" was

[1] Hercules.

hoisted on board by means of the fore spek tackle; a soft outside coating on the skin, called by the whalers gum, was scraped off with iron scrapers called gum knives, and the blubber taken off in long strips, yielding altogether about five or six hundredweight.

Narwhal.

The horn was on the left side of the nose, and on the right was a small one, not visible until after the blubber had been removed from the skull; it was about twelve inches long. The belly of the fish was of a beautiful pure white colour. Whalers, in a certain way, are most fastidious with regard to their eating, rarely partaking of anything, except birds, but what is brought out in the ship; the flesh of bears and seals is utterly discarded. This evening for tea we induced the steward to give us a portion of the narwhal's tail, which, boiled and steeped in vinegar, we found very palatable.

During the afternoon, seeing a large ground seal lying on the floe, I shouldered my rifle and attempted to stalk it, imitating as well as I was able, after the manner of the Esquimaux, the movements of the seal itself, crawling along the ice on my stomach and rolling my head about in a most uncomfortable manner, but consoling myself all the time for my very unpleasant mode of progression with the idea that I was doing the correct thing. My artifices, however, must have been very shallow and easily detected by my would-be prey, who disappeared in the water before I was within eighty yards. I was not sorry to get back to the ship, as the ice was cracking at every step I took in a far from pleasant way. If I had fallen through nothing could have saved me, as the strong current would quickly carry me under the ice, before any assistance could have reached me.

Thursday, June 19th.—This forenoon a boat was lowered after a narwhal in which I again officiated as boat-steerer, having Davy Smith as harpooneer. The "unie" was apparently frightened away, for we never saw it after leaving the ship. Whilst laying on our oars watching for it to rise, one of the crew declared he *heard* a whale, though a long distance off.

We all listened attentively, when Davy Smith also hearing it, though the sound was perfectly inaudible to my unpractised ears, we pulled vigorously in the direction from whence it appeared to come.[1] After pulling about a couple of miles through loose pack ice, we again lay on our oars and listened. Shortly the distinct blast of a whale was heard, this time by all of us, and away we went towards the spot indicated by the noise, as fast as five lusty rowers could propel us.

The ship, seeing that we had what the whalers call a "start," and knowing that we should not be pulling in such a vigorous manner if it was only

[1] Sir Edward Parry, in the account of his second voyage, mentions the extreme facility with which sound in cold weather is heard at a considerable distance, relating as an instance that one of his officers having occasion to send a man from the observatory to a distance exceeding one statute mile, had placed, for convenience, a second person half-way between to repeat his orders. This precaution, however, he found unnecessary, as he was without difficulty able to keep up a conversation with the man at the further station. The temperature at this time was —18°, and the weather nearly calm.

after a "unie," sent a couple of boats in the direction that we had taken, and shortly afterwards, having sighted the "fish" from the crow's nest, "all hands" were called, and all boats sent in chase.

And a fine chase this whale led us. For five hours did we pull *over* and *through* fields of ice without once being able to catch sight of our game, directed only by its blast. I have had many a hard day's work before in a boat in different parts of the world, which now appears to me as child's play compared with our exertions of to-day. The duty devolving on a boat steerer is by no means a sinecure when chasing a lively whale amongst heavy ice; at times the boat has to be forced between two floes, wide enough only to admit the boat, on which occasion she has to be sculled through by the steerer, and the incessant work of sweeping the boat round the numerous pieces of ice is most laborious. Shortly after 5 P.M. the bucket, to our no small relief, was observed hoisted at the mizen-top-gallant mast-head, as a signal for the boats to return, and I was by no means sorry to sweep the boat's head round in the direction of the ship. When once boats have been despatched in chase of fish, they are not at liberty to return until recalled by the bucket, or, if at a distance, by losing the fore-top-gallant sail. We had not pulled half-a-dozen strokes, before we heard the report of a gun, followed immediately by the cry of "A fall, a fall!" and we knew that our energetic and indefatigable mate was fast.

Repeating the cry, and sweeping the boats round, we pulled in the direction of the fast boat as lustily and cheerily as if we had not been at work, and at hard work, for the last five hours. Pushing the boat over streams of ice and between floes, and taking advantage of all open "leads," we were soon up to the fish, who had by this time received a second harpoon, and three or four fatal thrusts from a lance wielded by the experienced hand of Bob Gordon. From the immense jets of blood thrown up, contrasting strongly with the white ice upon which it fell, we knew she was near her end; and so it proved, for having anxiously waited with our deadly weapons in readiness for her reappearance, we commenced to haul in the lines—no easy task for seven or eight men in a boat, with a hundred-ton weight attached to the end. We had scarcely hauled in a couple of lines (200 fathoms) before the ship steamed up, and taking the lines on board and bringing it to the steam winch, the fish was dragged to the surface dead, amidst the vociferous and delighted cheers of all. She proved to be a perfect monster, worth at least five or six of those captured last Sunday. She will in all probability yield close upon twenty tons of oil. At eight we commenced to flinch, which operation was completed at about eleven o'clock.

We have been compelled to clear out some of our tanks in readiness to receive the blubber, and the coal with which they were filled, to the amount of about twenty tons, is heaped along the quarter deck. The state of the upper deck, from the agglutination of grease and coal dust, may therefore be better conceived than described. The voracity of the "mollies" swarming round the ship is perfectly astonishing; it is impossible to drive or frighten them away, though repeatedly struck at with the blubber knives and boat-hooks. The men at work on the whale alongside catch them easily with their hands, and will fling them back amongst their companions, who for the moment will rise affrighted on the wing, but only to settle again directly and resume the banquet. It is almost incredible to witness the enormous pieces of blubber that these birds can swallow; and they are so ravenous and greedy, that I have frequently seen them, after filling to repletion, swim a short distance away from the ship (being unable to fly in consequence of the amount which they had eaten), and, disgorging the contents of their stomachs, return again to the feast.

Walrus.

CHAPTER X.

MORE WHALES AT THE MIDDLE ICE.

FRIDAY, *June 20th.*—Off Cape Walter Bathurst. This forenoon, being close to the "Erik," we had a visit from Mr. Rickaby, a passenger on board that ship. From him we received the information that, after parting from us in Melville Bay, they went up to Dalrymple Island, a small islet between Saunders and Wolstenholme Islands, off Cape Athole; and that they were there visited by some Esquimaux, who came from the former island on sledges. They could speak no English, but by their manners they appeared as if they had recently had intercourse with white men, and this idea was corroborated by their having in their possession a couple of American government rifles, with the date 1864 stamped upon them, and which, from their clean and bright state, did not appear to have been long in their hands. From what Mr. Rickaby could learn, he was under the impression that these natives had very lately held communication with some white men to the northward, and that they—the white men—were either five days' journey (by sledge) from where they were (Dalrymple Island), or else five days from

Dalrymple Island.

Etah, the place from whence the Esquimaux had come. Referring to the ship, they held up two fingers, but whether they intended to convey the idea that there were two ships, or one ship with two masts, was uncertain.

I feel convinced that this must have some reference to the "Polaris," which vessel was schooner-rigged, and that she is in all probability either beset or wrecked not very far from Cape Alexander. It is difficult to account for the rifles being in their possession, as, being government property, it is most unlikely that they would be used as barter, unless in the last extremity. If the news regarding the "Polaris" that we heard at Lievely be true, it is more than probable that she has been abandoned by her crew and plundered by the natives. It is a great pity that the "Erik" did not attempt to obtain fuller and more complete information, as open water was reported as far as they could see to the northward.

In the afternoon, whilst steaming along the edge of a large floe, a bear was observed on the ice. Rifles were immediately in requisition. As we approached, Master Bruin, who had been watching us intently for some time, stood up on his hind legs and surveyed us with great attention; then, shaking his head, scampered off at full speed. A ball from my little rifle, however, stopped his career, passing through his skull, and killing him instantaneously. He was a young bear, measuring about seven feet from tip to tail; and the fact of his being clothed in his winter coat, makes his skin more valuable than it otherwise would be.

Saturday, June 21st.—The wind has literally been making sport of us all day; at one time blowing furiously from the N. W., the ship staggering under close-reefed topsails ; at another nearly calm, the sails furled, and the ship under steam ; and this has been going on for the last twenty-four hours. At noon to-day, during a violent gust, we made fast to an enormous field of ice, whose extent was not discernible from the nest, between Cape Liverpool and Cape Hay.

At 1 P.M. we made a further attempt to get up Lancaster Sound, but had to relinquish it in consequence of the severity of the gale. Orders were given to put the helm up, and we are now running back to our old fishing ground off the middle ice. I cannot get the " Polaris " out of my head ; I feel convinced, if we sought her, she would be found not far from the entrance to Smith s Sound. This morning, a large bear was seen swimming in the water, the ship's course was directed towards it, and a well-directed shot from the forecastle made it a prize. It was an enormous brute, measuring 9 ft. 6 in.

Sunday, June 22nd.—Yesterday the sun had obtained its greatest northern declination. I attempted to get its altitude at midnight, but it was unfortunately obscured by clouds.

" It is an ill wind that blows no one any good." Yesterday we were bewailing our hard fate at being unable to proceed up the Sound in consequence of the strong head winds which we encountered, compelling us to return to our old fishing station, expecting to hear, that whilst we had been uselessly expending our fuel in vain endeavours to get up the Sound, the remainder of the ships would have been actively employed amongst the whales. This morning, however, we observed from the mast head the *whole* of the whaling fleet beset and helpless, the recent strong winds having blown the ice out from Lancaster Sound. We are the only ship free, and, if we had fine weather, might take advantage of the utter helplessness of our consorts to pick up a few fish. The "Victor" joined us during the evening, having been detained in Melville Bay since we last parted company from her a fortnight ago.

Monday, June 23rd.—Another exciting and successful day. The wind, by noon, having subsided to a fresh breeze, and we being in a large bight formed by the ice, several whales were seen, "all hands" called, and the boats lowered away and in pursuit. In a very short time "a fall" was called, and then another, and by six o'clock we had a couple of fish alongside. One gave us a great deal of trouble, having taken to the ice, with upwards of a mile of line. It was some time before we could get at it, and then a difficult matter to kill it. One of the boats and its crew was literally drenched with blood; so much so, that when I saw it coming alongside, I imagined it was painted red, inquiring what ship it belonged to.

The other ships are all in sight, though, with the exception of the "Victor," completely beset. It must be very galling to their feelings to witness our success without being able to share in it. At midnight I obtained the meridian altitude of the sun, which was 6° 42′ 30″, or 6° 54 45″ corrected.

Wednesday, June 25th.—Last night was one of those beautiful evenings so frequently met with in these regions. The sun was shining brightly, whilst a light breeze scarcely ruffled the surface of the water, which resembled molten lead, with immense floes of ice floating on its bosom, whose sides were plainly reflected in the element on which they rested. The captain, who is possessed of remarkably quick eyesight, rendered doubly acute by constant observation and frequent use of the telescope, had just gone aloft, and I was admiring the tall, tapering spars of a whaler, so clearly and distinctly reflected in the water, about a quarter of a mile ahead of us, when I was awakened from my reverie by the skipper hailing in a low voice, ordering top-gallant sails and staysails to be set, and to turn ahead easily with the screw, though to put no fuel on the fire, saying at the same time he saw some fish about five miles ahead. We glided almost imperceptibly past the little vessel, whose shape and outline I had been recently admiring, before the people on board of her had discovered "what was in the wind." When fairly ahead "all hands" were called and all our boats quickly in pursuit. The light air had by this time entirely subsided, and the perfect calm which ensued rendered it extremely difficult for our boats to get within striking distance of the fish, who would both see and hear the approach of their enemies. I was up in the crow's nest with the captain (rather close quarters for two), watching the movements of the boats, and we could plainly perceive—and my companion took no pains to conceal his vexation—their utter failure to make a capture. At length, after three hours hard and untiring chasing, our energetic and hard-working mate succeeded in getting fast, and about an hour afterwards we heard the report of a gun from the direction which two of our boats had taken to the northward, and we plainly heard the faint cry of "A fall!" as it was wafted across the water. We on board instantly repeated the cry, shouting, "A fall, a fall!" But the skipper, with a most woe-begone expression of countenance, informed us curtly that it was no fall, that the fellow had missed. And so it proved. Poor Tom

Manson, in his excitement and eagerness to strike an enormous whale, imagining he would not get another chance, had attempted a long shot and had failed. The poor fellow's feelings may be easily imagined.

As all the fish had apparently been frightened away, it was deemed expedient to steam up towards the mate, whose fish had sunk to the bottom after being killed. The lines were brought to the steam winch, and we soon hove him up alongside.

He proved to be a fine whale, as he had also shown himself to be a very troublesome customer, having, after he had received four harpoons, chased one of our boats, whose crew had to pull for their lives to escape the terrific blows which were falling from the monster's tail, one blow being quite sufficient to entirely demolish a boat. The operation of flinching was not completed until 9·30 A.M., when everyone was sent to bed. After all it makes little difference in these regions what time one retires, day and night are so much alike.

The state of the ship is now almost indescribable. Cleanliness is decidedly not in the ascendant.

From the wheel aft to the taffrail, piled up in large stacks as high as the gunwale, is the whalebone taken from the recently caught fish; and from the wheel to the mainmast, on each side of the quarter-deck, are between twenty and thirty tons of coal, cleared out of the tanks in order to afford room for the blubber. The only *clear*, not *clean*, part of the ship is along each gangway, which must necessarily be kept so for the purpose of flinching and making-off. The captain says, "If we get another fish or twa we shall be in a fearsome mess." It seems impossible to be worse than we are. If we get any more fish we shall have to throw some of our coal overboard, as there is certainly no more room for any on our quarter-deck, the only available part of the ship; even the forehold has been cleared, and coal has been indiscriminately shovelled into the space vacated by the casks of provisions, which have been lashed to the bulwarks round the fore part of the upper deck.

Thursday, June 26th.—We are without doubt an extremely lucky ship. Last evening, whilst cruising in company with another whaler, a fine *fish* was seen lying alongside the edge of an extensive field of pack ice. We immediately despatched a couple of boats in pursuit, whilst our consort sent three.

After a short time the whale was seen to rise close to one of the latter, but, wonderful to relate, the harpooneer in charge, for some reason best known to himself, failed to take advantage, and the fish was not molested.

Again did the captain from the crow's nest hail to say that the other ship's boat would soon be fast; and again, to our great relief, was the opportunity missed. The next time the fish rose it was close to one of *our* boats, in which was Jemmy Gray, a sure and successful harpooneer, who, unlike his opposite number in the other boats, quickly fired and got fast.

"A fall" was called (it was about 10 P.M.), and it was amusing to witness "all hands" rushing up from below, scantily clothed, with their bundles in their hands, and tumbling, some of them head foremost, in their anxiety to get into the boats. I jumped into one, having the mate as harpooneer, and we all pulled vigorously towards the fast boat, which was entirely beset by ice, the other ship s boats slinking away on board their own vessel, thoroughly crest-fallen—and well they might be, as they had had two good chances of striking the fish if they had liked to avail themselves of the opportunity.

After laying on our oars for about ten minutes off the edge of the floe, the fish rose not far from us; we pulled quickly up and gave it another harpoon, two other boats coming up at the same time also firing their harpoons into the unfortunate beast. And now commenced the work of killing. Several times did the mate, with wonderful address, thrust the lance into the fish, pushing it in up to the handle (a distance of six or seven feet), we in the boat all this time being fully occupied in keeping close to our prey, but at the same time clear of its tail, which came down repeatedly with tremendous blows, drenching us with water. The poor brute soon commenced to eject blood from its blow-hole, with which we were copiously bespattered. At length the captain arrived on the scene of action, armed with a new American patent gun, loaded with an explosive dart, and two of these being fired into the whale, it turned over on its back dead, amidst our cheers. It was past 4 A.M. before we had finished flinching and retired to rest. I was delighted beyond measure, in the afternoon, when the captain announced his intention of going up Lancaster Sound, especially when a hint was thrown out of visiting Beechey Island and Port Leopold, and possibly going up Prince Regent's Inlet, as far as Fury Beach. With our worthy skipper to decide is to act, and we were soon steaming to the northward, with every prospect of being in a few hours on historical ground. There is a great deal of excitement in getting whales, but after all there is a great deal of sameness in this said whale fishery, which after a time gets monotonous. When once the fish has been killed and towed alongside, there is the same process to go through, which for a disinterested person soon loses its only charm, that of novelty.

The description of one whale being caught, and the subsequent operations, suffices for all.

Walrus Shooting.

CHAPTER XI.

LANCASTER SOUND AND BARROW'S STRAIT.

FRIDAY, *June 27th.*—Much to my vexation, this has been a thick, hazy day, with the clouds hanging so low as totally at times to obscure all glimpses of the land. At 8 A.M. we rounded Cape Liverpool, having a brisk south-easterly breeze in our favour. Passing a large glacier running down out of a deep gorge to the water's edge, off which several bergs, probably fragments from this very glacier, were aground, we came abreast of Cape Hay, which appeared steep and precipitous. At this place there is a large loomery, and as we passed we saw a countless number of these birds, both on the water and on the wing; but our time was too precious to allow us to stop for the purpose of bagging a few of these wild fowl, which afford such a pleasing variety to our everlasting beef; so the looms were allowed to remain unmolested, and in undisputed possession of their sterile and apparently inaccessible home.

About a mile to the westward of Cape Hay the water appeared to shoal to about the same distance off a point of land, off which were observed several small icebergs aground. We passed the place so rapidly that we were unable to obtain soundings. At about 1 P.M. we were off the Woollaston Islands, which are situated at the entrance of Navy Board Inlet. The clouds lifting for a short time, revealed to our view a party of about fifteen Esquimaux with their dog sledges on the land ice, probably come from Pond's Bay, for the purpose of "trocking."[1]

I was anxious to communicate with them, but "onward" was the order of the day, and they were soon out of sight astern.

The land ice extends in some places to about two miles off the shore, and directly across the mouth of Navy Board Inlet, so that at present the entire shore is unapproachable by boats. We have seen to-day a greater variety of the feathered denizens of these regions than we have hitherto seen in one day.

---

[1] To "trock" is synonymous with to barter, and is an expression always used by the Greenland Esquimaux.

Looms (*Uria Brunnichii*), as I have before stated, were observed in myriads; dovekies (*Uria grylle*) were likewise in great force; the rotges, or little auks (*Alca alle*), were also seen, but were left behind as we advanced up the Sound; the beautiful Arctic terns (*Sterna Arctica*) flew dreamily past the ships in flocks of ten or a dozen, whilst the delicate-looking ivory gull (*Larus eburneus*) and the graceful kittewake (*Larus tridactylus*) hovered in our wake. King eider ducks (*Somateria spectabilis*) were startled as they paddled along the edge of a floe, and flew rapidly past the ship, whilst our constant attendant and scavenger, the fulmar petrel (*Procellaria glacialis*) followed in our track, swooping down and devouring everything of an edible nature thrown overboard from the ship.

At six o'clock we made Cape Charles York, low-looking land, and apparently covered with some description of vegetation, it having a brownish-green sort of colour, which was pleasing to the eye, for we had seen nothing approaching verdure since we bade farewell to the bonnie hills of Scotland. Snow was lying in patches along the land, though the hills were still wearing their winter garb. The land ice, since leaving Navy Board Inlet, had gradually diminished in breadth, until it disappeared off Cape Charles York, from whence, however, as we proceeded to the westward it again increased in breadth, and we found the mouth of Admiralty Inlet impenetrably closed by a heavy and formidable barrier of ice.

This land ice appeared of great thickness and very hummocky, I should think quite twenty feet in thickness. At about seven, the sun shone out brightly, entirely dispersing the unpleasant mist by which we had been troubled the whole day but not sufficiently to dissipate the dense clouds which persistently hung over the land, concealing from my view much that I was naturally anxious to see.

Shoals of white whales were seen gambolling like porpoises along the floe edge, but they did not approach sufficiently near for me to get a shot at one. Sir Leopold McClintock succeeded in shooting one during his cruise in the "Fox," and speaks of its flesh as being palatable, and preferable to that of seal. By eleven we were off Cape Craufurd, the coast line, owing to the provoking clouds, being alone visible.

June 28th.—I was awakened at half-past three this morning by the cry of "a fall!" and the lowering away of the boats. Being surfeited with the continual capture of fish, and feeling rather drowsy, I determined to let the fish be taken without witnessing its death struggle, and turning over on the other tack, again courted sleep. But it was of no use, for in a few minutes the steward came in, and informed me of what I very well knew, that there was "a fall" I told him I was delighted to hear it, and again tried to get to sleep. After a lapse of a few minutes, the skipper bounced into my cabin, saying in a voice loud enough to wake the dead, "Are you awake, captain?" When I informed him of the important fact that I was wide awake, and had been so for some time, he told me that Davy Smith was fast to a fine whale. Seeing it was hopeless to expect any more rest, I turned out, dressed, and went on deck in time to see the third harpoon fired and the fish killed.

It was the same old story, so I shall refrain from saying anything about it. At five o'clock we sat down to breakfast No. 1, and at half-past eight, after the flinching was over, to No. 2. The fish had evidently been struck before, having a scar on its back and a healed wound in its flank, which appeared to be caused by a harpoon. This makes up over 100 tons—not bad luck, considering we are only eight weeks from Dundee to-day. What would not such men as McClintock, Osborn, or Richards do in command of a vessel of this description? Not two months from England, and we are off Prince Regent Inlet, with no ice in sight to the westward. I verily believe we could easily this year make the north-west passage and come out safely with our ship by Behring's Straits. The weather is most tantalizing, a heavy fog hanging over the land, utterly screening it from our view. At six this morning I obtained good sights for double altitude and chronometer, and also the meridian altitude at noon. We had supposed, since eight o'clock, that we were steaming up Prince Regent Inlet, and the captain went to bed under that belief. I felt convinced we were steering nearly a due west course, and told the mate so, the compass being so sluggish, in consequence of our proximity to the magnetic pole, as to be rendered nearly useless. At about four the fog suddenly lifted, exposing to our view land along the *port* beam, and precipitous cliffs immediately ahead, and within half a mile! It seemed a providential escape! My own predictions were fully realized; we had been steering to the westward, and the bold, steep land ahead was Leopold Island.

At six o'clock the sun broke out bright and strong, so much so as to make it unpleasantly warm walking the deck; but the abominable fog still clung to the land, and, as if sorry for indulging us with a glimpse of Leopold Island, again wrapped its impervious mantle around it.

7.30 P.M.—The fog cleared up, exposing to our

view the east land of North Somerset. No ice was in sight, but a blink was observed ahead. The land along the coast appears steep and precipitous, though sloping down in places towards the water's edge. It is lofty table land, and its height I should estimate to be from three to four hundred feet above the level of the sea. The easternmost point of Leopold Island is very abrupt, and from the southward appears to be an overhanging cliff; so also Cape Clarence. We passed close to Port Leopold, where Sir James Ross wintered in '48; it seems a snug winter harbour, but, though I looked long and attentively, I failed to see any indication of the depôt supposed to exist there. Cape Seppings, which forms the southern extreme of Port Leopold, slopes down to the water, and is hardly deserving the name of a cape. The east point of Leopold Island is a very prominent feature, and if named would be of some use.

9 P. M.—Again we are enveloped in thick fog, and I am thankful for the brief glimpse of land I have had. A short distance from the north of Elwin Bay we came up to the land ice, stretching across the inlet; fires were eased down, and we are again under canvas. I am in hopes we may soon be among fish, in which case I shall try and get the skiff, and make a running survey of the

coast of North Somerset from, if possible, Fury Point to Leopold Island; it would not take me more than four or five days if the weather were fine, and would be to me a most interesting and delightful occupation—in fact, any change would be most agreeable. The captain has certainly a most wonderful constitution for sleep; during the last forty hours he has been over thirty in his bed; the only time he was out was when the whale he caught was being flinched; he is already regretting coming here, and thinks we should have done much better in the middle ice. Man is never content—"the more he has, the more he wants."

The total absence of ice to the northward will certainly render it almost impossible to fish should a breeze spring up from that direction, for it would cause such a sea to get up, as to make it dangerous and almost useless to lower boats. We are now feeling the motion of the ship more than we have done since entering Baffin s Bay.

Sunday, June 29th.—A most disagreeable day, a fresh breeze blowing straight up the inlet, raising a nasty tumble of a sea, reminding me in an unpleasant manner of the Atlantic. We have also our usual accompaniment, a villainous mist which, freezing on the ropes, causes a shower of ice every time we tack, the rigging being perfectly encrusted with a hoary coat. We have been cruising backwards and forwards along the land ice, which extends across the inlet from North Somerset to the opposite shore, nothing being visible in consequence of the thick weather which prevails. Taking advantage of the utter inability to fish, everyone, with the exception of the regular watch, has devoted himself to the drowsy god, and I have been left all day in sole possession of the cabin and the greasy spot between the main mast and the winch, which I have selected as my only place for exercise. Two other whalers were observed at intervals during the day. I am much afraid that the ice will prevent us from ad vancing further up the inlet, and the captain does not like the idea of my leaving the ship, as he does not know when he may be compelled to go out again. If we see no fish, we shall certainly return to our old fishing ground, the middle ice, between Cape Byam Martin and Cape Graham Moore.

Monday, June 30th.—A lovely bright, clear morning. The sun, shining brilliantly over the snow-clad table-land on each side of the inlet, owing to the peculiar transparency of the atmosphere, presented that lustrous effulgence rarely witnessed out of these regions. Hardly a breath of wind ruffled the surface of the clear, pellucid water on which we rested, whilst from the southward, stretching across from Batty Bay, where Capt. Kennedy, in the "Prince Albert," wintered in '51, to Port Bowen, where Sir Edward Parry wintered in '24, was the land ice, an effectual barrier to our further progress up the inlet. Suddenly all was activity and commotion on board; whales were seen, and all boats lowered and sent away in pursuit, the "Camperdown" following our example, and in less than a couple of minutes no less than sixteen boats dotted the water, destroying the effect of the hitherto unbroken and fairy-like scene. There was no time to be wasted, for the "Camperdown's" crew appeared to be the leading one. At about eight, we observed one of her boats to be fast, and ours was, therefore, recalled and hoisted up after being away about two hours; our consolation—a very poor one—being that it was only a small fish, "no larger," as one of the harpooneers informed me, "than a sardine."

Tuesday, July 1st.—2.30 A.M.—Just returned from a most successful foray against the huge mysticeti. Having fortified the inner man, I shall attempt to detail the proceedings that have occupied our time for the last seventeen hours. At about half-past nine yesterday forenoon some whales were seen from the crow's nest, about five miles off, and a couple of boats were sent away to try their luck. Shortly more whales were seen, and all boats were ordered away in chase. We were at this time about abreast of Leopold Harbour, steaming out of the inlet on our way to the middle ice outside Lancaster Sound. Immediately whales were seen the engines were, of course, stopped. One of the boat-steerers having severely injured his foot the day before, the mate offered the post to me, an offer I gladly accepted, as a means of passing the time and indulging in the excitement of whaling. We pulled steadily for at least six miles without seeing anything, the ship by that time being well down astern. It was a calm, lovely day, with bright sunshine. Soon the blast of a fish was seen, followed by several others; but being a clear day, our approach, when near, was always observed, and down they would go, with a tremendous splash of their huge tails. At about twelve, Jemmy Grey—I have quite adopted the custom on board the "Arctic," and always allude to the harpooneers by their Christian names—got fast to a fish; leaving a couple of boats to assist him, the remainder dispersed in various directions, chasing whales, all of which seemed of a small size. It was nearly two when I saw a heavy blast some distance ahead, accompanied by a small one, which, in duty bound, I reported to my harpooneer, who pronounced it to be a "monstrous big fish with its sucker" (young one). Away we went in chase, another boat in company. Having judged, from observing the way in which she was heading, which way she would rise, we lay on our oars and waited for her. Our plan was skilfully ordered, for she rose close to Harky Hunter, who, pulling quietly up, got his harpoon in, we all yelling at the top of our voices, "A fall, a fall!" But though so easily struck, she was not so easily killed, and proved herself to be a most troublesome and awkward customer. After about three quarters of an hour's hard pulling, chasing her round and round the fast boat, the men being so fatigued as nearly to drop at their oars, we succeeded in getting close up and giving her another harpoon, immediately on receipt of which she flew off at a terrific rate, towing the two boats at least six knots an hour, and taking the line out with such velocity that water had continually to be thrown over the lines and over the bollard head, round which three turns had been taken, the bows of the boat being enveloped in smoke caused by the friction. We had a very near shave one time of being taken down altogether. The fish having stopped to blow for a few minutes, the line, which, from the extreme friction, had burnt a deep scar round the bollard, had cooled and adhered to the wood. The fish suddenly took it into her head to go straight down: the line would not render, the bows of the boat were dragged under water, and the water came rushing in over the harpooneer. This saved us, for the water lubricating the line, allowed it to render, and the boat righted, though not before a large quantity of water had been shipped. We should, indeed, have been placed in a most dangerous and unpleasant predicament, for had the line not rendered, nothing could have saved the boat from being taken down, and our chance of escape would have been very small. The other fast boat was some distance from us, and they would have thought twice before cutting their line, and so losing a valuable fish, to come to our assistance, the other boats being miles away. Our stroke oar and line manager, a powerfully-built Shetland man, standing about six feet two inches, commonly called "Big Johnnie," was unable to swim, so that altogether we should have been in a pretty pickle. It was five o'clock before any boat came to our assistance, when five more harpoons were buried in the monster's flesh, and several lances plunged in, but all, apparently, to no avail,—the brute would not die. Three rockets were also fired into the unfortunate animal, that clung to life with such tenacity. Eventually the ship came up and took the lines from one of the boats on board, and yet, singular to relate, the fish actually towed the ship and seven boats at the rate of three miles an hour, the water in which we were towed being coloured with blood , but her furious struggles gradually weakened, loss of blood and the powerful efforts she had made to free herself, necessarily caused great exhaustion, and at nine o'clock a boat was able to come up, and, firing a rocket, succeeded in giving the *coup de grâce* with a lance, and she expired amidst the cheers of all hands. We had been over six hours fast to this fish, during which time we had been towed a distance of upwards of fifteen miles, Cape Hurd, which was not in sight when we left the ship, being distinctly visible not many miles on our starboard bow, the fish having headed up the sound. I can safely say, if any one asks me how I went to Barrow Straits, that I was towed there by a whale. No sooner was the fish killed than we were able to turn our attention to others round us, and three more small ones were struck. I was not sorry to see the bucket up on board, and at half-past eleven P.M. we all returned to the ship, hungry and tired, having been away, and therefore famishing, for fourteen hours. I shall very soon reach my berth, and have no doubt of enjoying a good night's rest;—four fish in one day is not a bad day's work. The little "sucker" that was with our big fish was seen in its company for about half an hour, coming up to blow every time its mother rose for that purpose, and then suddenly disappeared after the mother had given some unusually violent convulsion with its tail and fins. I am told that a whale having its young one in company will, when struck, invariably kill it if she gets a chance, which accounts for the disappearance of our small one.

3 P.M.—Everything very quiet; hardly a soul on board the ship awake except myself; the ship running up the inlet before a light north-west wind, apparently her own master, for no one is on deck, and the ship steering herself. Flinching was completed about eight o'clock, and everybody being thoroughly done up, went to their beds, where they are to be allowed to remain undisturbed until four o'clock. The fish appear to know this, for in the forenoon there were a number of them playing round the ship in all directions, and we actually struck one with our fore foot, which was laying ahead of the ship, apparently without the effect of frightening it away, for it rose shortly after close to us; the captain, however, very properly, will not call the men up until they have had their sleep. Some time ago I asked the captain if he had ever known or heard of a case in which a ship had come into collision with a whale, and he answered in the negative. Such an event cannot, therefore, be of frequent occurrence, though it is mentioned by the historian

of Frobisher's third voyage to the north-west in 1578, that one of his ships, the "Salamander," "being under both her corses and bonets, happened to strike a great whale with her full stemme with such a blow that the ship stoode still, and stirred neither forward nor backward. The whale thereat made a great and ugly noyse, and cast up his body and taile, and so went under water." A few days after they fell in with a dead whale, supposed to have been the one struck by the "Salamander." The only part of this story which appears doubtful is that in which it says the whale made a noise, as no sound except their blast is ever heard from them when struck by a harpoon; the simple striking of a whale with the stem of a ship, and such a ship as Frobisher had in his expedition, would not in my opinion cause its death, as I know from experience the enormous amount of killing some of them require.

The men were turned out at four, having an hour given them to collect their scattered senses and to get some breakfast, dinner, or tea (it is impossible to say by which name this meal ought to be called), and were then set to work "making off" the blubber of the last five fish that were caught. Words cannot describe the filthy state of the whale-ship. Casks of provisions are lashed to the bulwarks round the fore part of the upper deck, so as to make room in the fore hold for coal, which has been indiscriminately shovelled into the space vacated. The whole of the quarter-deck is covered with coal and whalebone; sawdust is sprinkled everywhere to prevent the men slipping in the slimy matter with which the deck is coated, and this, combined with the grease and coal-dust, clogs and sticks to the soles of our boots, rendering it quite impossible to avoid bringing it down into the cabin. When speaking of whales you must always call them "fish," nothing else in whaling parlance coming under that designation. This I found out the other day on seeing the blast of a narwhal, and exclaiming, "There is a fish." "No, sir," said Tom, the second mate, "that's nae fish, only a unie." We are cruising off and on between Leopold Island and Cape York, at the entrance of Prince Regent's Inlet. Now that we are ready for fish, they have made themselves scarce.

Wednesday, July 2nd.—Flinching was not completed until eight o'clock this morning, when the watch was set. The state of irregularity in which we live is positively delicious. I never know when I am to have anything to eat, nor even when I am to get any sleep. On some days we breakfast at 5 A. M., on others at 5 P. M., and the remaining meals are served in the same charming proportion of irregularity. We are now again going up Prince Regent's Inlet, the land on each side barely discernible on account of the mist.

Thursday, July 3rd.—This morning I was awakened early by the captain coming into my cabin to tell me that the "Esquimaux" and the "Narwhal" were both coming up the inlet; and about eight, as we were sitting down to breakfast, we received a visit from Captain McLellan, of the latter ship, from whom we learnt with surprise that the two ships had only been released from their icy bondage the previous Saturday, having been beset ever since we last saw them off Cape Byam Martin, and had been drifted to the southward of Port Bowen. The "Narwhal" had fared very badly, having received some very severe nips; indeed, at one time there was very little prospect of saving the ship, the floes having been squeezed and pressed up on each side as high as the gunwale, so that they were enabled to walk straight out on the ice. Provisions and men's bags had been landed on the floe, together with the boats, and during four days they were hourly expecting to witness the total destruction of their ship. All this time was of course a dead loss to them as regards fishing. There was a very nasty tumble of a sea last night, which has now subsided, though, in consequence of a dense fog, nothing is distinguishable.

Looms.

CHAPTER XII.

PORT LEOPOLD.

FRIDAY *evening, July 4th.*—At three o'clock yesterday afternoon, in consequence of our inability to fish, owing to the wind and high sea, the captain determined to put into Port Leopold to wait for fine weather and to fill up with water. I was delighted when I heard the orders given to get up steam, and for the course to be altered to the northward. We steered close along the eastern shore of North Somerset, the land rising steep and precipitous, and giving one almost the idea of passing along the side of a huge fortified wall, with embrasures about a third of the way from the bottom, and regular buttresses along its face. It seems difficult to account for the peculiar formation of these buttresses, which are arranged so methodically; but I should suppose that it is to be attributed to the action of the weather, on the soft sandstone of which the cliff is composed. At about five o'clock we rounded Cape Seppings, and shortly afterwards made fast to the land-ice, which stretched across the mouth of the harbour, just inside the two points. On the right hand, as we entered, was Whaler Point, where we observed the skeleton framework of a house, a boat turned bottom up, and numerous casks scattered about and covering a large extent of ground.

My thoughts went back to 1848 and 1849, when Sir James Ross wintered in this very place with the "Enterprise" and "Investigator." The house was erected by his orders, and the provisions were left as a depôt for the relief of the ill-fated crews of the "Erebus" and "Terror." Sir James Ross entered the harbour of Port Leopold on the 11th of September, 1848, having found a greater quantity of ice in Barrow Strait than had ever before been seen at that period of the season. Had the two ships not got into port on that day, it would have been impossible to have done so any day afterwards, the main pack, during the night, having closed the land, and completely sealed the mouth of the harbour. Sir James Ross then proceeded to land a good supply of provisions upon Whaler

PORT LEOPOLD.

Page 178.

Point, in which service the steam launch, brought out by the "Investigator," proved of great value, conveying a large cargo herself and towing two deeply-laden cutters through the sheets of young ice which covered the harbour. Without the aid of steam a boat could not have moved her own length. Sir James also caused a house to be built of spare spars and covered with housing cloths, and on his departure he left, besides the provisions, the "Investigator's" steam launch, lengthened seven feet, and made capable of conveying the whole of Sir John Franklin's party to the whale-ships. Sir James Ross's expedition was in Port Leopold from September 11th, 1848, until August 28th, 1849. In this desolate spot three of the greatest of our modern Arctic navigators thus found shelter for nearly twelve months. Besides Sir James himself, the discoverer of the North Magnetic Pole, there were M'Clure, the future discoverer of a North-west Passage, and McClintock, the discoverer of the fate of Franklin. On September 21st, 1851, Mr. Kennedy, of the "Prince Albert," landed and passed the night in the steam launch, marching thence, with a sledge, to his winter quarters further south, in Batty Bay. He was there again on July 27th, 1852, while on his long sledge journey with Lieut. Bellot, round North Somerset, when he deposited a record.

The place was afterwards again visited by McClintock on August 19th, 1858, when in command of the "Fox"—the final expedition sent out in search of Sir John Franklin. He found that the ice had been pressed in upon the low shingle point, forcing the launch up before it, and leaving her broadside on to the beach, with both bows stove in, and in want of considerable repairs; but the means were all at hand for executing them. He left a whale-boat, and added a record of his proceedings to the many that had accumulated here during the ten years between 1848 and 1858. In that interval several exploring ships, employed in the search for Franklin, had touched at Port Leopold as they passed up or down Barrow Strait or Prince Regent Inlet. McClintock found the harbour entirely clear of ice.

As I was looking towards Whaler Point, and thinking over its past history, my reverie was suddenly disturbed by the captain, with whom I went on shore. On landing, I found a tin cylinder, red with rust, lying on the beach, which I rightly guessed to contain the records. I picked it up and strolled away, eventually returning on board. On opening the cylinder I found a roll of paper neatly marled over, but thoroughly saturated with water, and, on unwrapping the outside sheet, there was another as neatly put up as the first, on which I could just make out, in Sir Leopold McClintock's handwriting:—

"Records." "1849."
"Fox."
"19th August, 1858."
"F. L. McClintock."

Inside were the older records, but so thoroughly soaked as to render it almost impossible to open them without tearing, the writing being nearly illegible. I resolved, after some consideration, to dry and preserve the records, and to bring them home, leaving a notice to that effect in the tin cylinder, which I caused to be soldered up, and secured with wire to the centre support of the house erected by Sir James Ross. My chief reason for coming to this decision was the risk of these interesting relics being destroyed or thrown away by future crews of whalers who may land here. The documents, which have now been delivered over to the Hydrographer of the Admiralty, in the order of their dates, commence with three signed by Sir James Ross, and dated August, 1849. Here is an extract from them:—

"It is earnestly desired that any persons (not in absolute distress) who may find the stores and provisions landed here by H.M. ships Enterprise' and 'Investigator,' will leave them undisturbed."

Next was a record, dated August 13th, 1850, left by Mr. Saunders of H.M.S. "North Star;" and another dated August 28th, 1850, left by Captain Forsyth, R.N., of Lady Franklin's searching vessel, "Prince Albert." The next was a record deposited by Mr. Kennedy, commanding Lady Franklin's searching vessel, "Prince Albert," during 1851-52. She had wintered in Batty Bay, on the west side of Prince Regent Inlet, and her commander, with a travelling party, visited Port Leopold on July 27th, 1852. The last record is that deposited by Sir Leopold McClintock on August 19th, 1858; from which the following is an extract:—

"After taking on board some coke, lemon juice, and soap, we shall proceed this evening upon our voyage. The provisions in casks and canisters appear to be sound. I leave a whale boat here as a resource for any retreat party from this vessel, in consequence of the bad state of the launch, and shall not leave a boat at Fury Beach."

The provisions on shore were still in a wonderful state of preservation, the biscuits being as good as when baked, and the tobacco and chocolate quite fresh after twenty-five years. It seems desirable that, when any of these stores are used by the whalers, a correct account of what has been taken from this depôt of public property should be furnished to the Admiralty.

There were five graves at the head of the harbour, of one officer and four men belonging to the "Enterprise" and "Investigator," who died during the winter of 1848-49. The officer was Mr. Henry Mathias, the assistant surgeon of the "Enterprise," who died on the 15th of June, 1849, aged 27, of consumption, a disease which had been deeply rooted in his constitution before he left England. Sir James Ross spoke of him as "a promising young man, of great amiability of disposition, universally beloved and regretted." The men were James Gray, David Jenkins, and Edward Binskin, able seamen of the "Enterprise," and William Grundy and Thomas Coombs of the "Investigator." At the foot of the grave of Coombs, who belonged to the carpenter's crew, there was a bottle containing a document with the following touching but rather quaint inscription:—

"Near this spot lay the remains of Thomas Coombs (late belonging to the carpenter's crew of Her Majesty's ship 'Investigator'), who died on board that ship on the 27th day of October, 1848, after a lingering illness of three months, which he bore with Christian fortitude. And I sincerely hope, should any Christian fall in with *this*, that he will leave his body rest in peace and undisturbed, and oblige his late chum and messmate, Charles Harris, A.B."

All the inscriptions were neatly engraved on plates of brass, which were nailed on the head boards. The boards were of two-inch wood, and, though bleached white from the effects of weather, were in a good state of preservation. They were in the usual shape of a grave stone.

In the evening of the same day, taking my gun, and accompanied by the mate and steward, I went away to get some eider duck, flocks of which were flying about. We succeeded in shooting eleven king ducks, four long-tailed ducks, and several looms and dovekeys. The long-tailed duck (*Fuligula glacialis*) is much smaller than the eider, being about the size of a widgeon, and having much the same sort of plumage, but with long tail feathers ending in a point like the boatswain bird. Having got a good bag, I was anxious to obtain a view by climbing the hill on the south side of the harbour, which is about 600 feet high.[1] I induced my companions to come with

[1] Sir John Ross climbed to the top of this hill, on the 2nd of September, 1832, at the time when he was vainly attempting to get out into Baffin's Bay, and was obliged to return and

me, though we found the ascent to be more difficult than it appeared, being very steep. The side of the hill is covered with layers of a slaty stone, having the appearance of closely packed tiles. From the summit we had a glorious view up Barrow's Strait, which, however, seemed to be blocked with ice. The opposite shore was distinctly visible, and we could plainly see, what I imagined to be Beechey Island and its surrounding land. It being 3 A.M., the morning sun was

---

winter at Fury Beach (see "Narrative of his Second Voyage, p. 665). But it is not generally known that he then put a tin case under a heap of stones, which was never found; although the officers and crews of the "Enterprise" and "Investigator" were here for so many months in 1848-9. In the tin case Sir John deposited a copy of verses which he had composed for the occasion, and in 1852 he gave a copy of them to Mr. John Barrow, who has kindly forwarded it to me. A poetical effusion, at such a temperature, by so old a navigator, is worthy of being preserved.

> "Far as the eye can reach and all around
> Is one vast icy solitude profound.
> On snow-clad ground, in silent stillness, sleep
> The weary crew; no soothing vapours steep
> The rocks with freshness, not an herb is there,
> Nor shrub, nor bush,—but, desolate and bare,
> It seems as if these regions, by the will
> Of Heaven transfix'd, had all at once stood still;
> And the proud waves, beneath the fatal blow,
> Had spread into a field of lifeless snow."
>
> JOHN ROSS, Sept. 1832.

shining brightly, causing the land to stand out in clear and bold relief. Our descent was much more rapid than our laborious climb up the hill. Following the mate's example, we sat down on the frozen snow, and then slid the whole way to the bottom—a quick mode of progression, though desperately fatal to one's clothing. We returned on board at four in the morning, pretty well worn out by our day's exertions.

The "Esquimaux" whaler had meanwhile arrived in Port Leopold, and made fast to the land ice close to the "Arctic." On the 4th I again went on shore, and had a good look round. The house, of which only the framework is standing, is about thirty-five feet long by twenty-five feet, and it could easily and expeditiously, by means of a few old sails, be made habitable. The flag-staff had been blown down, and was lying uninjured on the shingle. The launch was a perfect wreck, but the whale boat appeared sound, and could easily be made seaworthy. There were about two hundred and thirty casks of different sizes and descriptions lying about in all directions, suggesting the idea that they had originally been very carelessly landed. But their positions may probably be attributed to the bears, whose inquisitive habits are well known, and the marks of teeth and claws were visible on several casks. Much of the biscuit was in a decayed state, though some was perfectly good. The whole of the tinned meats were in good order, as also were the sugar, chocolate, and tobacco; tea and raisins quite perished; peas in fair, and flour in a very good state. The blankets were almost as good as new, and I was much astonished at the remarkable state of preservation of the rope, which had been exposed to the weather for twenty-five years. It was bleached perfectly white, resembling Manilla rope, but on opening the strands the heart was quite sound. This speaks volumes for the preservative qualities of an Arctic climate. There were two large heaps of coke near the depôt.

Port Leopold is, without exception, the most barren and dreary looking spot it was ever my lot to behold; no signs of verdure anywhere, nothing but sandstone, snow, and ice. Occasionally, in some little sheltered crevice, I came across a bit of green moss or pretty little pink flowers growing in small patches, but not in sufficient quantity to relieve the desolate sterility which everywhere prevailed. I brought away with me, as a memento of my visit, a rusty sailor's knife, with a buckhorn handle, having a hole through the haft for a shackle, such as is always carried by a man-of-war's man, which I picked up close to the shed. I wonder if its former owner is still alive!

In the afternoon I went on shore with the artificial horizon, and obtained a good set of sights, and a round of angles; but smoke issuing from the ship's funnel warned me of our approaching departure, and I had to hurry on board. We steamed out of Port Leopold at 3.30 P.M., and once more began the look-out for whales.

Saturday, July 5th.—The "Arctic" is decidedly a lucky ship, and manages to tumble on the fish in a most remarkable manner. Last night at twelve o'clock, as I was enjoying my last stroll, preparatory to turning in, up and down my greasy promenade, the captain suddenly hailed from the nest, saying he saw a number of fish about ten miles ahead, ordering all hands to be turned out, and get their coffee in readiness for a start, which took place at about half-past twelve. A boat-steerer being invalided, Davy Dewchers, a loose harpooneer, requested my services; and away we all went. After pulling for about six miles, a beautiful morning, with a bright sun, and the water as smooth as glass, we came up to the fish, and before long heard a shot fired and "A fall!" cried, and Bob Gordon was fast to an enormous whale. At about six o'clock a fish was close to our boat, which after a few starts we were fortunate enough to strike, and in our turn called "A fall! a fall!" She took out a little over two hundred fathoms of line, when the mate coming up to our assistance fired a second harpoon, and succeeded in killing her in about an hour's time. Hauling our lines in and flaking them away, we got up to our fish, and lashing its two fins together, round the tail, up to the bows of our boat, and in this manner, stern foremost, pulled towards the ship, which was nearly hull down in the distance. We were all naturally much elated at our success, as it was my first fast fish; and it was also our harpooneer's first paying fish, he never before having harpooned one over six feet, no whale under that size being "payable." Of all the uninteresting and toilsome duties connected with whaling, there is nothing, I thirm, can beat our morning's work, which was the laborious duty of towing a dead whale for more than eight hours. We did not get alongside the ship until eleven o'clock this forenoon. On arriving we found that they had just succeeded in killing the first whale struck, which, like the one we got a few days ago, had been most amazingly troublesome, having towed the ship and five boats some distance. They were nearly nine hours killing her, but she proved to be a perfect monster, larger than any we have got. Tom Webster, the second mate, was also lucky enough to get a couple of small whales. Four fish again at a fall! though the one large one is worth about twice as much as the three others. We commenced flinching a little before twelve, which lasted until 7.20 P.M. At eight all hands were sent to bed, and I am again apparently in sole possession of the ship. As it is now nearly 10.30 P.M., and I have not been in bed since yesterday morning, and only have had three hours sleep during the last sixty-four, I shall follow the good example set by everyone, and also retire.

Sunday, July 6th.—I am becoming rapidly conscious of the fact that I am gradually falling into the insidious clutches of that enemy which has for so long successfully attacked and held sway over my shipmates and everything on board the "Arctic,"—that irresistible and encroaching foe—dirt!

Yesterday, when I returned from our long chase after the whales, I was in a most deplorable state, for the boats are, if such a thing can be possible, in a more filthy condition than the ship, and I find that my clothes are all impregnated with a conglomeration of grease, tar, and coal dust, and other things too numerous to mention. We were all considerably refreshed by a good night's rest, and I felt very loth to turn out this morning when called by the steward, especially as he had not then had time to light the fire, and there was a fresh cold wind blowing from the westward, the effect of which, in consequence of the spanker being set—which acts as a famous, though unpleasant, wind-sail—was distinctly felt in our cabins. It required no small amount of resolution to plunge from a nice warm bed into water at freezing temperature. The sun was shining brightly, but the wind was so cold that I noticed a canvas jacket, which had gone through a nominal process of washing, and was hung in the rigging to dry, in less than twenty minutes frozen as hard as a board, while its lower edge was adorned with a beautiful fringe of icicles. Frequently, while away in the whale boats, I have conversations with the men, generally relative to the Arctic Regions, and I was extremely gratified to learn that in case of a Government expedition leaving England next year with which I was connected, that more than a third of the ship's company have professed themselves willing to accompany me; and should such a desirable event take place, I shall always know where to get volunteers, and could pick out at least twenty strong, hard-working hands out of this crew that, once set going, would never rest until they had reached the North Pole. At half-past eleven this forenoon more whales were seen, and all hands sent in pursuit, and were not long away before "a fall" was called. This makes our twentieth whale and about one hundred and seventy tons of oil. I much fear that my chances of going north will diminish in proportion to the rapidity with which the fish are caught. A hundred tons more and the ship will be full.

Long-tailed Ducks.

Meeting of "Arctic" and "Ravenscraig" off Cape Craufurd.

CHAPTER XIII.

THE POLARIS EXPEDITION.

MONDAY, *July 7th.*—"One never knows what the morrow may bring forth" is a saying frequently verified, and seldom in a more surprising manner than by the events which have happened during the last twenty-four hours. I had just gone to bed last night when the captain came down and informed me that the "Ravenscraig" was in sight, steaming up the Sound. We were at this time off Admiralty Inlet on our way to the middle ice. As the information was not very interesting to me, I endeavoured to get to sleep, when the engines were stopped, and I heard the captain hailing. Suddenly I caught the sound of the word "Polaris" and "survivors," which caused me to jump speedily out of bed; but before I could dress, a messenger had been sent down to tell me that a portion of the crew of the "Polaris" had been picked up by the "Ravenscraig," and that our captain had gone on board. Hastily dressing, I lost no time in following him, and on the quarter-deck of the "Ravenscraig" was introduced to Captain Buddington and Dr. Bessels of the ill-fated exploring ship. The news that we heard at Lievely, relative to a part of the crew of the "Polaris" having been picked up off the coast of Labrador, was now corroborated, eighteen[1] of them having been drifted off on a floe; and the

---

[1] Namely:—

‡1. Tyson (2nd captain).
*2. F. Meyer (asst. to Dr. Bessels).
‡3. John Heron.
‡4. J. W. C. Kruger.
‡5. W. Jackson.
*6. W. Lindemann.
‡7. Peter Johnson.
*8. F. Antinn.
†9. G. W. Linguist.
*10. Fred. Jamka.
11. Hans.
12. Hans' wife.
13. Joe.
14. Hannah.
15. Tuomy.
16. Succi.
17. Augustina.
18. Tobias.

(11–18) † Esquimaux.

remaining fourteen[1] having passed the winter of 1872-73 near their ship, to the northward of Cape Alexander, in the entrance of Smith Sound. The vessel herself was run on shore. On the 4th of June the party of fourteen left for the south in a couple of boats of their own construction, flat-bottomed scows made from the bulwarks and other timber, and were picked up by the "Ravenscraig"[2] 25 miles S.E. of Cape York on the 23rd, having by that time only two or three days' fuel left, but in other respects they were well supplied. During the boat voyage they encountered no special dangers or hardships. The greatest inconvenience they experienced was the want of tobacco. They consoled themselves by smoking tea, which they say was a very fair substitute. As Captain

---

[1] Namely :—

‡1. Capt. Buddington.	‡8. Hayes.
*2. Dr. Bessels.	*9. Siemens.
‡3. Capt. Chester (1st mate).	*10. Henry.
§4. Mr. Morton (2nd mate).	‖11. Campbell.
*5. Mr. Schumann (eng.).	‡12. Bryant (asst. astron.).
‡6. Mr. Odell (2nd eng.).	‡13.
‡7. Coffin (carpenter).	‡14.

* Germans, 8 ; † Esquimaux, 8 ; ‡ Americans, 14 ;
§ Irish, 1 ; ‖ Scotch, 1 :—Total, 32.

[2] The "Ravenscraig" was nipped by the ice at the time, and the sketch, drawn on the spot by Dr. Soutter, the surgeon on board, shows the crew of the "Ravenscraig" hauling the boat of the crew of the "Polaris" over the ice towards the vessel. *See next page.*

Adams was anxious to take some of the crew of the "Polaris" on board the "Arctic," it was eventually arranged that Dr. Bessels, Mr. Chester the first mate, Mr. Schumann the engineer, and four men should come with us, the others remaining on board the "Ravenscraig." It was 6 A. M. before all arrangements were concluded, and we bade farewell to the "Ravenscraig," which vessel proceeded up the sound, whilst we steamed out to the middle ice.

"Ravenscraig" off Cape York, and boat of "Polaris."

The expedition of the "Polaris," under the command of Charles Francis Hall, will always be remarkable for having proved the navigability of the strait leading from Smith Sound to the north. At present the "Polaris" has reached a higher northern latitude than any other ship on record.

It is to be hoped that this will not continue long to be the case, and that she will only point the way to the vast unknown area. Meanwhile every detail respecting the voyage of the "Polaris" is important, both as regards the discoveries and the inadequate means with which they were effected.

The commander of the expedition, Charles F. Hall, was a native of Cincinnati, where he was engaged as editor of a newspaper called the "Daily Penny Press;" and he had formerly been apprenticed to a blacksmith. He is described as a man with a compact, vigorous frame, and with a firm expression of countenance; but he had no advantages from education, and was unacquainted with nautical astronomy He was thus in no sense a seaman, but rather an enthusiastic leader, depending on others to navigate his vessel and to render his discoveries useful. He possessed, however, one great advantage. His two previous expeditions had thoroughly acclimatized him, and given him a complete knowledge of Esquimaux life. The men who accompanied him were also badly chosen. Buddington was an old whaling captain, without any interest in the undertaking; and Tyson was a man of the same stamp. Chester, the mate, was a good seaman and excellent harpooneer, but one who had merely shipped from the inducement of high pay. Dr. Bessels, a former student of Heidelberg, who had served in one of the German Arctic Expeditions, and in the Prussian army during the invasion of France, was the only man of scientific attainments in the ship, and the only man, besides Hall, who felt any enthusiasm for the objects of the voyage. Yet while such a man as Buddington received 120 dollars, the pay of Dr. Bessels was at first only 75 dollars a month. He made the American Government ashamed of their niggardly offer by requesting them to appoint some one to assist him in his astronomical observations, to whom he proposed to transfer his salary. Though they did not accede to his request, they raised his pay to 100 dollars a month. He is a great friend of Dr. Petermann, and is convinced that the theoretical geographer of Gotha will change his opinion respecting the best route for reaching the pole when he has read the report of his countryman Bessels. Altogether it was an ill-assorted company, without zeal for discovery, without discipline or control, and in which every man considered himself as good as his neighbour. Hall and Bessels were the only two among them who really desired to reach the pole.

The most striking fact connected with the voyage is that the "Polaris," in August, 1871, went from Cape Shackleton to her extreme northern point up Smith Sound in 82° 16′ N.[1] in five days, and even then she was stopped merely by loose floes through which a powerful vessel like the "Arctic" could easily have forced a passage. I was indeed informed that the "Polaris" was stopped by a very insignificant stream of ice which, in addition to its offering no real obstruction, had a clear lead through into open water, with a magnificent water sky as far as could be seen to the northward. Hall was most reluctant to turn back, but being no sailor and having no experience in ice navigation, he thought he had no alternative but to follow the advice of his sailing master, Captain Buddington. This old whaling skipper, fearing that if they persevered they might be unable to retrace their steps, advised a retrograde movement, and thus ended all further attempts to reach the North Pole. The floes met with up Smith Sound were not of a heavy description, and seldom exceeded five feet in thickness, so that we may infer that they were formed in one winter. No icebergs of any size were seen to the north of 80° Dr. Bessels informs me that, at Newman Bay, a place about eighteen miles to the northward of their winter quarters, the ice appeared heavier and more extensive than that further south, though it

---

[1] This latitude was estimated, and was not by observation.

was all drifting to the south and west. Two or three of the fields he estimated at a mile in length, with hummocks from ten to twelve feet high, but all the ice seen from their winter quarters appeared to be only of one year's growth.

From their highest point, in 82° 16′ N., they saw land to the north and west, which they estimated to extend as far north as 84° It also appears that, while the south coast of Greenland is gradually subsiding, the north is rising. As a proof of this, Dr. Bessels brought away a sea-water shrimp which he had taken out of a fresh-water pond, thirty-eight feet above the sea level; and he picked up some marine shells at an elevation of 1,200 feet above the sea. A mussel shell, in 81° 45′, was found at a height of 1,600 feet.

The coast of Greenland extends only as far as 82° 30′ N., whence it trends away to the eastward, and there seems to be little doubt of its insularity. It is steep and precipitous, and is free from land ice; while the shores of Grinnell Land, on the opposite side, appear to be low and shelving, and have fast ice attached to them. The extreme northern point of Grinnell Land appeared to reach about the same latitude as the north-west point of Greenland, or a little further to the north, and then to trend away to the westward, leaving a channel between these two countries and a northern land visible at a distance of sixty miles. This would place the latter in about 84° N.

There was a current flowing down the strait from the north; and a small piece of drift wood was picked up in 82° N. It was a bit of pine wood, about a foot long. In the same latitude Mr. Chester, the mate, discovered a great deal of drift wood, which, as he did not attach any value to it, was used as fuel. One piece was over four feet in length. The Esquimaux Jem, who is really a west coast native, though now settled with the Etah people round the mouth of Smith Sound, told me that plenty of wood came from the north, and was washed up along the shore of Grinnell Land. It must be borne on the waves of a great polar sea from the coast of Siberia. The most interesting discovery connected with oceanic movements was, that the tidal waves from the north and south meet at Cape Fraser, on the west coast of Grinnell Land. This was fully demonstrated by the drift of the ship and by tidal observations. To the south of Cape Fraser the flood tide makes to the north, whilst to the north it flows south. The rise and fall during spring tides was about five and a-half feet, and during the neaps about two feet. No agitation of the water was noticed off Cape Fraser caused by the meeting of the two waves, for the ice would effectually prevent anything of the sort. But to the south of Cape Fraser the tide rose to a greater height during the night, as is the case along the coast of Greenland; whereas to the north of Cape Fraser there was no perceptible difference between the day and night tides.

Vestiges of the presence of man on the verge of the unknown region are proofs of its being habitable, and therefore of the presence of open water and much animal life. Dr. Bessels saw traces of Esquimaux as far north as 82°, in which parallel he picked up, lying on the beach, a couple of ribs of the walrus which had been used as sledge runners, and a small piece of wood that had formed part of the back of a sledge. An old bone knife handle was also found, and the remains of a summer encampment, consisting of three circles of stones for keeping tents in position, with spaces left for the entrances.

As regards the work of previous American expeditions in Smith Sound, I was informed that all the coast line laid down by Hayes, and the "open polar sea" of Kane, are quite imaginary. Morton, the steward, who is said to have discovered the wide immeasurable ocean, was on board the "Polaris," and we brought him home in the "Arctic." He is an Irishman from Dublin, and a very good man, and he took the mild chaff that was levelled at him about his famous "open polar sea" very good-humouredly. Cape Constitution of Kane has been determined to be about fifty miles south of the position formerly assigned to it by that explorer, and the entire coast line must be placed considerably further to the eastward. Dr. Bessels satisfied himself of the existence of the United States Sound of Hayes, on the west coast of Smith Sound, and is impressed with the belief that a passage may be made through it into the Polar Sea, which would establish the insularity of Grinnell Land. He is also of opinion that the Lady Franklin Bay of Hayes is another opening to the westward. One of his reasons for this belief is, that the large ice fields met with at the highest latitude reached by the "Polaris," though drifting south, were never seen to the south of Lady Franklin Bay; and it is his impression that they go up that sound or strait.

From her furthest point the "Polaris" was drifted to the southward, in consequence of the prevailing north-east winds, until she was able to get into winter quarters in 81° 38′ N., in a sort of harbour formed by a small iceberg, on the east side of the strait. Hall died in November, 1871, soon after returning from a short autumn travelling excursion; and my messmates of the "Polaris" all appeared to be of one opinion, namely, that had he lived, the expedition would have been a complete success. Dr. Bessels informed me that his leader's death was caused by apoplexy, materially assisted by his own want of caution when returning very cold from his sledge journey. He arrived on board the ship much chilled at about three in the afternoon, and immediately went below and had some hot coffee, without taking off his furs. At about six o'clock he was taken ill, and died in a few days, being quite insensible for some time before his end. He was a teetotaller, and was much annoyed at seeing others drink. There can be no doubt that Hall was an enthusiast, and that his whole soul was wrapped up in the noble desire to achieve greatness. He has found a last resting place in the midst of his discoveries.

The winter quarters of the "Polaris," in 81° 38′, are the most northern position in which civilized man has ever wintered; and all details respecting the temperature and the amount of animal life are consequently most interesting. The lowest temperature registered was —48° Fahr., with very little wind blowing at the time. The prevailing winds were from the north-east. The fall of snow during the two winters passed by the "Polaris" up Smith Sound was remarkably small, the heaviest snow storm occurring in the month of June, and that was not of any extraordinary amount. In the latitude of their winter quarters musk oxen were met with, and twenty-six were shot. Foxes and lemmings were also seen, but other animals were comparatively scarce, and only one bear was seen during the whole year. Narwhal and walrus were not seen to the north of 79°, but seals were obtained up to the extreme point in 82° 16′ They were of three kinds, namely, the common Greenland seal, the ground seal, and the fetid seal. The bladder or hooded seal was not met with. On the western side it was stated by the Etah Esquimaux, that Ellesmere Land abounded with musk oxen; and, judging from the configuration of Grinnell Land, the same abundance of animal life is to be found there also. The birds all disappeared during the winter, though ptarmigan and a species of snipe made their appearance early in the spring; and in the summer all the genera found in other parts of the arctic regions were abundant. With the exception of a salmon seen in a fresh-water lake not far from the beach, no fish were met with. The contents of the stomachs of the seals they caught were found to consist of shrimps and other small shell fish. Dr. Bessels used the dredge on several occasions, but owing to the ice, he could seldom do so at a greater depth than eighteen or twenty fathoms, the results being generally unimportant, and with the exception of a few shrimps and other *crustacea*, nothing of interest was obtained. No less than fifteen species of plants, five of which were grasses, were collected by the doctor at their highest latitude, on which the musk oxen must subsist. He gave me four specimens of the *flora* of 82° N., the names of which will be found in the list of plants in the Appendix, to which Dr. Hooker has kindly added an explanatory note. Mr. Chester presented me with a fossil from the Silurian limestone of that high latitude, which is also referred to in the Appendix. Dr. Bessels made a fair collection of insects, principally flies and beetles, two or three butterflies and mosquitos; and birds of seventeen different kinds were shot in 82°, including two sabine gulls and an Iceland snipe.

Dr. Bessels disagrees with those who assert that total darkness prevails during the arctic mid-winter. He showed me his observations, and a sketch made in 81° 38′ (their winter quarters), from which it appears that, on the 21st of December, they had not less than six hours of twilight, which rose at noon to an altitude of 10° During the first winter they lost twenty-five out of the sixty dogs they had on board, most of them dying in violent fits. This disaster was attributed to the carelessness of the steward, who threw scraps of salt meat into the lee scuppers, which were eagerly devoured by the dogs. A depôt was established at the winter quarters in 81° 38′ N., consisting of 1,000 lbs. of pemmican and other provisions, besides guns and ammunition. With the exception of poor Captain Hall not a man died, and although many were well advanced in years, all returned in perfect health.

The "Polaris" was released from her winter quarters in June, 1872, and after sending out a few parties with unimportant results, her head was turned homewards. She was then beset and drifted out into Baffin's Bay, where she sustained a severe nip in the following October. On the 15th of that month it was blowing a heavy gale of wind, and the ship was closely beset. Provisions and boats were landed on the ice to provide for the worst. Suddenly the ship broke out and flew before the wind at the rate of ten or eleven knots, leaving nineteen hands on the floe, men, women and children, with the boats and provisions. The weather was very thick and the ship was leaking badly, so she was run on shore near Lyttelton Island, just inside Smith Sound. Dr. Bessels attributes the wonderful set of the ice experienced by those who were left on the floe, as well as the drift of the "Resolute," "Fox," and the vessels under De Haven in 1851, entirely to the wind and not to the current. Out of the nineteen souls thus separated from their ships, nine were Esquimaux (two men, two women, and five children), and the remainder were Germans and Americans. Mrs. Hans, the wife of the Esquimaux who was with Kane and Hayes, was safely delivered of a son in 82° N. The little fellow has, in all probability, the most northern birthplace of any human being living.

During the second winter, near Lyttelton Island, the remaining crew of the "Polaris" received much willing assistance from the Esquimaux of the Etah settlement. Dr. Bessels visited Port Foulke, the winter quarters of Hayes in 1860-61, and on examining the grave of Mr. Sontag, the astronomer of that expedition, he found that the Esquimaux had dug up the remains for the sake of possessing themselves of the wood of the coffin. The bones were scattered in all directions. These he collected and re-buried. On inquiring for the stores which Hayes had left behind him in the observatory, he was told by the natives that that edifice had been blown up by the powder left in it, causing the death of five Esquimaux. They had probably broken into the place for the sake of plunder, and owing to their carelessness with regard to fire, the powder stored there had ignited. Dr. Bessels also ascended the *mer de glâce* of Hayes, and reached an altitude of 4,181 feet, at a distance of twenty-eight miles from the coast.

The "Polaris" expedition was certainly fortunate in securing the services of a man so full of resources as Dr. Bessels. He is ever ready to invent some substitute if an instrument is broken, and is of a most ingenious turn of mind. Amongst other things which illustrate his inventive genius I may mention his contrivance for amusing the crew with a Christmas tree during the first winter. The tree was made of tissue paper, and placed in their observatory on the shore of Polaris Bay, and must have been a decided success. On a shelf overhead he placed two large basins, one containing grog and the other punch. From these basins he led a couple of india-rubber tubes which were concealed by the foliage of the tree, a mouthpiece only projecting from the branches. Then, by pressing a valve with his foot, the doctor was able to give the men, as they entered with their pannikins, either grog or punch as they desired, much to their surprise, and to their no small satisfaction. It is often very amusing to hear the chaffing discussions which take place between Dr. Bessels and Mr. Chester, the mate, who seem to have a high opinion of each other. One day the latter was boasting of his cooking qualifications, when the doctor reminded him of a certain *scouse* he had made for the party, and which even the dogs were unable to eat, the ravenous propensities of the Esquimaux dogs being proverbial.

Dr. Bessels considers that no expedition should go north without some Esquimaux for hunting and dog driving. He gave me the names of a man and his wife belonging to Etah, in whom he has the greatest confidence, the woman being of use in making boots and skin clothing, and the man being an excellent driver. The doctor having served in the German Arctic Expeditions, has had experience of the two routes for Arctic exploration. He is now decidedly of opinion that Smith Sound is the best, if not the only route by which the North Pole can be reached.

American discovery ship "Polaris."

Chapter XIV

AGAIN IN BAFFIN'S BAY.

JULY 10th.—We are now on our old fishing ground off Cape Graham Moore. I am beginning to lose all interest in the fishing, and shall be only too glad when we have succeeded in filling up, and our head is pointed to the southward. I shall be disappointed if the "Arctic" is not the first to carry home the intelligence of the rescue of the remainder of the "Polaris" crew. This has been another of our delightfully irregular days. Last night, at about half-past ten, fish were seen and the boats sent away, and after a little dodging about, Tom Mason got fast to a fine whale; it was no sooner alongside than the boat was again sent away in chase, and at about 10 A.M. Davy Smith got a splendid chance, and fired his harpoon right into the unfortunate monster. Mr. Chester and myself went away in the dingy with the rocket gun; and it was fortunate that we did so, for when the fish next rose it was in the midst of loose pack ice, through which no boat could possibly have penetrated. We determined on attempting to reach our prey on the ice, and leaving the boat, we commenced a rather ticklish and perilous journey, as the ice was of a most rotten and treacherous nature. On several occasions we were nearly going through, and the boy who accompanied us and was carrying the ammunition fell in, and was only saved by throwing himself forward. Unfortunately he lost all the ammunition; however, we had loaded the gun before landing, so that one shot was preserved. Chester had the gun, while I was armed with a lance. After walking rather more than half a mile, which took us nearly an hour to accomplish, we came up to our game, which was lying in a little hole that it had made for itself through the thin ice. Chester fired, and with effect, for almost immediately the dart exploded the fish turned over and expired. If we had not gone on the ice with the gun it must have escaped us, having only been struck with one harpoon, and we could not possibly have lanced it from the ice. Our journey back was as hazardous and difficult as the one made to the fish. However, it has been a good day's work, as the two whales caught are very large.

Friday, July 11*th.*—We had our first meal at half-past twelve (A. M.), and it was amusing to hear every one as they came into the cabin and looked at the clock, inquiring "if it was half-past twelve *to-morrow* morning, or half-past twelve *last* night." In the afternoon a bear was seen, and Chester and myself immediately started in the dingy in chase. Fully carrying out the character for curiosity attributed to these animals, Master Bruin actually ran along the edge of the floe towards us, apparently to ascertain who the audacious creature was that dared to approach and disturb him during his meals. A shot from my little rifle striking him full in the face brought him down, but I was compelled to fire twice more before life was extinct. There was a huge piece of whale tongue lying on the ice, which he had picked up in the water and deposited in its present position. It must have weighed at least a couple of hundredweight; it was quite fresh, and was most probably a piece of *kreng* thrown overboard yesterday. Bruin measured over seven feet eight inches. Shortly afterwards I shot a large ground seal, but the beast tumbled off the ice, sank, and was lost. The last two days have been thick and foggy, much to our annoyance as it renders all search after whales impossible. The arrival of the doctor and party from the "Polaris" has, indeed, been a god-send to me. Time was beginning to hang very heavily, for this whaling business becomes after a time extremely monotonous, and is relieved only by the actual capture of a fish. The subsequent operations begin to get tedious and uninteresting, usually terminating in dirt, weariness, and discomfort.

July 13th.—We have unfortunately broken our hydrometer, and therefore are unable for a time to observe the specific gravity of the sea-water, which, with our other observations, we have been regularly taking every two hours; but the doctor is most philosophical with regard to his loss, and I do not doubt that in a short time he will devise some instrument by which we may continue our observations.

Monday, July 14th.—When I rose this morning it was actually raining; the first rainfall we have had since leaving the Atlantic. I am afraid I did not greet it as an old friend, for it materially assisted in making the upper deck, if such a thing were possible, in a more filthy state than it was before. The doctor fully confirmed his reputation for ingenuity by constructing this forenoon a hydrometer made out of a small bottle, a quill, and a little of my mercury. It is a decided success.

In consequence of a thick fog which prevailed in the morning, the ship was made fast to a floe, and the men employed in making off the blubber from the last whale caught. Clearing up in the afternoon, steam was raised, and we are now proceeding in the direction of Pond's Bay, intending to fish along the edge of the floe.

Tuesday, July 15th.—Being thick and foggy this morning, we again made fast to a large floe, and filled with water obtained from small pools on the surface. With my artificial horizon I was able to get sights for latitude and longitude. Last night we saw the "Eric" and "Polynia" off Pond's Inlet, but the land-ice extended so far out, and the middle-ice was in such close proximity, that the captain would not venture on making fast for fear of being beset, so we are again cruising about under canvas. Whilst alongside the floe, I got a cast with the deep-sea lead, but obtained no soundings in 200 fathoms. The temperature of the air was 41·9°, at the surface 35·5°, and at extreme depth 32° In the afternoon as I was talking with Dr. Bessels, all my other messmates being asleep, I was told there was a bear on the ice. I hurried on deck with my rifle, and went away in the dingy with the doctor and a couple of hands. Fearing that he might escape us by running shorewards (that is, away from the water),

I landed on the ice and walked round him. Bruin, however, for a wonder, was not of a curious turn of mind, for when I had approached to within 120 yards he jumped into the water. I ran as hard as I could to the edge of the floe, and just catching sight of him as he turned round to have a look at me, sent a bullet through his head, killing him on the spot. He was a fair-sized brute, measuring seven feet eight inches. Ever since last Sunday we have had the most villainous foggy weather; occasionally the sun will break out, dispersing the mist, but only for a very short time, when we are again enveloped in fog. It is most tantalizing, as it is impossible to keep a look-out for fish, for not even Sam Weller's "double magnifying glasses of hextra power" could pierce the dense mist which surrounds us. We are also amongst much ice, and it is impossible to find our way out; however, it is, thanks to calm weather, loose and open, though there is no saying how long it may remain so.

Thursday, July 17th.—A beautiful clear, calm day. We have steamed away from the obnoxious ice-floes, and now resume our weary occupation of looking for fish. At noon I got a cast with the deep-sea lead, having had good sights, so as to fix the ship's position, but with the usual result—no bottom at 200 fathoms. The temperature of the air was 40°; of the surface, 38·2°; at 200 fathoms, 30·5° By way of experiment I attached an empty bottle, well corked, and with a good coating of sealing-wax, to the line just above the lead. When hauled to the surface, the cork was found to be completely driven into the bottle by the pressure of the air, and the bottle was consequently filled. The doctor has discovered amongst his relics a solar thermometer, the ball of which he painted over with lamp-black, so that, in addition to our other observations, we are also able to register the temperature in the sun. At 2 P. M. it was as high as 118·2° He has now taken possession of the second berth in the captain's cabin, by which arrangement I again become sole possessor of my own bed. He generally retires about 2 P.M., and rises again between 12 and 1, so that he records all the observations between 2 A.M. and noon, and I take them from 2 P.M. to midnight. He served during the war as surgeon in a Prussian regiment, and interested me very much to-day by relating some of his experiences during that campaign.

Friday, July 18th.—A fine day—a real arctic day—which, when it comes in these regions, is truly enjoyable, especially when it comes after thick, foggy weather; it carries with it an exhilarating influence rarely met with elsewhere. Though our thermometer on deck only registered six degrees above freezing-point, the warmth of the sun is so great that one feels inclined, to use a vulgar expression, to peel. It is a day on which gloves, mittens, and mufflers are discarded, and comforters laid aside. Even at midnight yesterday the thermometer showed sixty-four degrees in the sun. It was long past one before I retired to rest this morning, so pleasant was my solitary walk while enjoying the beauties of an arctic night, forgetting for a while, until brought to my senses by slipping on the greasy deck, that I was really and truly on board a Dundee whaler. Not a ripple ruffled the smooth water, which had the appearance of a polished mirror, reflecting everything that rested on its surface, and magnifying the sea-birds that lay or paddled about on its bosom to such a size, that they resembled more seals' heads or whales' crowns, than our constant attendants and scavengers, the greedy fulmar petrels. Loose ice lay dotted about in small clusters or patches, and it would not require a great stretch of imagination to represent these patches of ice as sheep huddled together in numerous flocks, with here and there single animals scattered about, straying from their fellows, in a delightfully smooth and rich pasturage, the greenish colour of the water materially assisting the illusion. A few small bergs, raising their crests a short distance above the horizon, might easily be taken for cottages or shepherds' huts, and thus complete the landscape, while to add to the beauty of this charming picture, there was a bright sun, about five degrees above the horizon, yet not so lustrous but that it was possible to look at it without pain to the eyes. Overhead there was a glorious sky, what is generally termed a "mackerel sky," the blue and white of which were gradually blended together as they descended towards the horizon, having rich tints of various bright hues, in regular and separate layers, directly opposite and round the sun. It was truly a delightful night, but it is utterly impossible for me to describe it so as to do any justice to the scene. To-day there is a light breeze from the southward, which makes the chance of getting fish a little better. We have never, since our arrival on the fishing-ground, been so long idle, and it is all the more tantalizing, when we only require four or five more whales to complete our cargo, and enable us to spread our wings for our homeward flight. Our restless captain cannot, as I anticipated, endure this protracted period of idleness, and the engines are now throbbing away full speed, and our head pointed in the direction of Cape Byam Martin, whence we shall again proceed up the sound and try our luck in

Admiralty and Prince Regent's Inlets. With fortune on our side we may very reasonably expect to fill our ship in ten or fourteen days' time, and then start for home.

Pond's Bay.

Chapter XV

ADMIRALTY INLET.

J*ULY* 19*th.*—It is surprising how very popular an expedition of discovery towards the North Pole appears to be with every one. Yesterday the mate was inquiring, in case a government expedition started, whether there would be any chance of his being appointed as ice-master. I have met no man in these ships better qualified for such a duty, or more competent for such a situation; young, powerful, a good walker, an excellent shot,—these are a few of the qualities he possesses, joined to a most daring and enterprising character. If I had a ship going north he is a man I should like to have. One of the crew of the "Polaris," a German, also came to me, saying he had heard that there was a chance of an expedition going next year, and that he was very desirous of joining it. He is a fine, intelligent man, of whom they all speak most highly, and was one of the two who saw the most northern land, estimated in eighty-four degrees. To-day has been one of the finest, and decidedly the warmest, day we have had since leaving England. Temperature as high as fifty-seven degrees. At eight o'clock we were off Cape Byam Martin, since which time (it is now past midnight) I have been taking angles, bearings, and altitudes, and working out the results. Dr. Bessels was kindly sketching the land for me as far as Cape Liverpool, after which I lost his services, in consequence of the necessity of his going to bed at 2 P.M. It is difficult for me now to do any writing in the evenings; for since the arrival of the Americans, the game of euchre has come into fashion, and the table is taken up by it. The game began this evening at 7.30 and did not cease until 11.

I am quite surprised at the inaccuracy of our charts. The whole trend of the coast-line is incorrect. Cape Liverpool would be more properly called Point Liverpool, being merely a long promontory jutting out into the sea. One distinct and high headland between Byam Martin and Fanshawe is altogether ignored on the chart, and this I have named Cape Sherard Osborn. I counted twelve large glaciers between Capes Fanshawe and Hay; but including small or embryo glaciers, there are fully

THE "BARTLE FRERE" GLACIER.

Page 222.

four-and-twenty. The largest one is directly behind Cape Liverpool, and is about five or six miles in breadth. None had the appearance of being discharging glaciers, and the depth of this one did not appear to be very considerable. I have named them the "Bartle Frere Glaciers," after the President of the Royal Geographical Society. At eight o'clock, being off Navy Board Inlet, a couple of boats were sent away in chase of a whale, and the captain kindly letting me have the dingy, I landed, accompanied by Chester and the two doctors, on the north-west point, and succeeded in getting some sights, though the altitude was so low as to compel me to lie flat down and close to my horizon. We afterwards collected a few botanical and geological specimens, and shot a few birds, the doctor having promised me some lessons in skinning them.

Sunday, July 20th.—Sunday always seems an eventful day with us, and this has been unusually so. At seven this morning, off the mouth of Admiralty Inlet, we met the "Narwhal," the captain of which kindly sent us a couple of "Dundee Advertisers," one of which had an account of the rescue of the remaining portion of the crew of the "Polaris." Luck still remains with us, for having, in company with the "Narwhal," made fast to the land-ice stretching across the inlet, a fish was seen close to the latter vessel, when boats were instantly sent away in pursuit. Before long a chance presented itself, the harpooneer of the "Narwhal's" boat fired, and the cry of "A fall! a fall!" told us they were fast. By some unaccountable accident, probably the harpoon drew, or it was not securely embedded in the blubber; however, the fish was again loose, and rose not far from one of our boats, two of which were "on the bran." We saw Harky rise to his gun, and heard the report, but instead of the cry of "A fall!" and the hoisting of the jack, to our chagrin we perceived that he had missed. The skipper concealed his feelings wonderfully well, and went below; but strange to say, the fish had not been frightened, and again rose not far from Harky, who pulled up, and this time got fast. The whale was soon killed and alongside. It was not a very large one, but still it is another *fish*, and gives us altogether 210 tons on board.

On arriving at the land-ice, several parties of Esquimaux came down to us, and the ship has the whole day been besieged by them—dirty, unkempt-looking people, both men and women. They arrived in sledges drawn by from nine to twelve dogs each.

Altogether there were seven sledges, bringing about twenty-five men, women, and children. With the exception of a few foxes' skins and walrus tusks they had little to barter, though that did not prevent their asking for everything they saw, and the more that was given them the more they wanted. I have received by no means a favourable impression of these Pond's Bay natives, for from that place they come. They seem to me about the lowest specimens of humanity I have

Esquimaux dogs.

ever come across, even including the Solomon Islanders.

One man picked up the carcase of a loom that the doctor had been skinning, and seemed to relish it amazingly, tearing off the raw flesh with his teeth; observing me watching him intently, he offered me a piece, but I need not say the tempting morsel was declined. In consequence of its being a cloudy day I was unable, much to my disappointment, to get sights. In the afternoon we suffered for our warm temperature by a perfect drenching downpour of good heavy rain, which lasted many hours. The natives sought shelter at once. They appear to have a great abhorrence of water—at any rate, I can vouch that none is ever used for cleansing purposes.

I am sorry to say we are again without our hydrometer, it having been accidentally dropped overboard while Hayes was drawing water. Last night when we were away in the dingy we pulled between the two westernmost islands of the Wollaston group, in a crevice between the rocks, on one of which we found the nest of a little snow-flake, containing the young birds and one egg, of which we possessed ourselves in a cruel and heartless manner.

Monday, July 21st.—ELWYN INLET.—As I anticipated, it came on to blow very hard from the northward at midnight, causing a heavy sea, which broke up our ice and enforced our speedy departure, the captain putting in here for shelter: a snug little haven, where we are laying comfortably made fast to a land floe. The last we saw of our friends, the Esquimaux, they were busy packing their sledges and harnessing their dogs ready for a start, as the ice was cracking in all directions. When we came into this place it seemed literally swarming with narwhals. Very shortly after sending our boats away, we had secured seven, off which we shall get about a ton of blubber. Unfortunately they are all young and females, so, consequently, have no horn, excepting one that had an appendage about two feet in length, scarcely worth mentioning. I had a long shot at one flying through the water, but though I struck him the harpoon failed to get fast. All the afternoon I have been hard at work taking sights and angles, both from the ship and on shore, which will enable me to make a fair plan of the harbour. I succeeded in obtaining soundings from the ship in 190 fathoms soft muddy bottom, temperature 29·5°, surface 34°, air 37°; and also collected several botanical and geological specimens.

Tuesday, July 22nd.—An unpleasant, damp, foggy day, at times so thick as to render it impossible to distinguish the end of the bowsprit. This thick weather is a great drawback as regards taking sights, the fogs, especially at this time of the year and next month, being so prevalent as to make a clear day quite an exception.

We left Elwyn Inlet about ten, and soon afterwards the captain put his head into my cabin with the welcome intelligence that the "Tay" was in sight, and the dingy was despatched for our letters and newspapers. We are now fishing along the floe across Admiralty Inlet, but keeping well over to the west side. Occasionally, when the fog is not so dense, we catch sight of the "Eric," "Tay," and "Narwhal," all the ships keeping close together, and in what they call the thick water, in which fish are only to be seen. We have seen one or two whales during the day, and have lowered our boats after them. The mate, however, got fast to a narwhal, which, although a female, had a horn three feet in length—but without success. According to the accounts given in the English papers by the "Polaris" survivors of the dimensions of the floe on which they drifted down to the coast of Labrador, its circumference, when they started, was five miles, which was reduced to twenty yards in diameter before they abandoned it. Both the doctor and Chester assure me, the former having actually measured it, that a mile and a-quarter in circumference was the outside; so that if its decrease in size is exaggerated in the same proportion—in reality it will not have been much reduced. I employed myself this morning in examining, sorting, and stowing away my geological specimens, and find that the land on the east side of Navy Board Inlet is composed of Silurian limestone similar to Cape Hay. In addition to this limestone there are several pieces of drift, some off large boulders, probably transported by ice, and which consist of gneiss, mica-schist, and granitic gneiss. The land about Elwyn and Admiralty Inlets is composed of the same Silurian limestone, which rises up precipitously from the sea to a height of from 1,200 to 1,500 feet. Its striations are most distinct, though very irregular, in places being nearly horizontal, in others dipping towards the northward and westward from 6° to 10° During the remainder of the day I have been working out my sights.

In the evening it was suddenly discovered that there was very little water left in the ship, so we are now hauled to the wind, beating up towards our old harbour, Elwyn Inlet, the weather still continuing as thick as pea-soup. I am not sorry we are going back, as, if the sun shows itself to-morrow, I shall be able to complete a very fair plan of the anchorage.

Wednesday, July 23rd.—At midnight last night we made fast to the land ice in Elwyn Inlet, and immediately had all the men and boats away watering ship, the water, pure and fresh, running down the sides of the cliff into our casks. So expeditiously was this duty performed, that by 5 A.M. we had received over twenty tons of water, and taken our departure from the inlet, without giving me a chance of getting any observations, the weather still remaining thick and foggy, the wind having fallen, or, to use a whaling phrase, "having taken off." We pushed through the loose brashy ice, and made fast to the main pack across Admiralty Inlet, where we are now patiently waiting, with a couple of boats "on the bran," for any fish that may be unwise enough to approach within sight. The ice is from five to seven feet in thickness, but with a change of wind it will quickly break up.

Saturday, July 26th.—A wet, thick, foggy, and unpleasant day. We are cruising about along the floe edge, still in Admiralty Inlet; but the weather is decidedly against our fishing, irrespective of a fresh north-westerly breeze, which raises an unpleasant tumble of a sea. This effectually puts a stop to any work that I may attempt below, for no sooner are my papers and instruments out than down they all come, for their eternal game of euchre. From shortly after breakfast until past midnight is this game going on, so I am compelled to resort to the upper deck.

Sunday, July 27th.—Still enveloped in this persistent and villainous fog. I begin to think we are destined never more to see either land or sun. The wind has subsided and the sea has gone down, but the thick weather seems to exert a most depressing influence on all on board, though for different reasons: on my shipmates, because of their utter inability to fish; and on myself, because I am unable either to get observations or to see the land. We imagine ourselves to-day to be off Cape Craufurd, a point the position of which I am especially anxious to determine. Our pile of coal on the quarter-deck is diminishing in a most alarming manner, which gives me a great deal of uneasiness, though the captain is constantly raising my spirits by telling me that I am destined to see much more than any one that ever came out in a discovery ship.

Iceberg.

Chapter XVI.

PRINCE REGENT'S INLET.

MONDAY, *July* 28*th*. — Again in Prince Regent's Inlet. At midnight last night the fog suddenly cleared off—rapid changes being very common in these regions—and found us abreast of Cape Craufurd, distant about eight miles. Steam was immediately raised, and a course shaped for this inlet. It was beautifully calm until 1 P.M., when a strong breeze sprang up from the westward, causing the spray to wash clean over our decks; but by 4 P.M. we were steaming through an immense quantity of loose pack ice, which gave us smooth water; and thinking fish might be about, the screw was stopped, and the ship put under canvas. The entire coast line, from Cape Craufurd to Cape York, is incorrectly delineated on the chart; I was able to get a fair notion of the trend of the land, but unfortunately we were too far off to make a really accurate survey. I obtained sights at 4.30 A.M., 5.30 A.M., 6.40 A.M., 9.30 A.M., and noon; and with the true bearings obtained at the same time, shall be able to give the principal headlands.

Tuesday, July 29th.—This has been a day full of adventure and excitement, one of those days one reads of in story books treating of the whale fishery. Last night a fine narwhal was harpooned, with a horn rather over nine feet in length After breakfast I went away to officiate as harpooneer in a boat "on the bran," but the water was so smooth, and everything so silent and quiet, that we were unable to approach the "wily unie" sufficiently near to effect its capture. Before returning on board I hazarded a long shot at one, but although it was hit, and we all thought we were fast, he got away I had just completed a double altitude when the captain sent down to say he saw some bears on the floe. Putting my books on one side I ran on deck, and getting the dingy, went away with the two doctors and Chester. We pulled for some time without seeing anything, and were just on the point of giving it up and returning to the ship, when I espied "Brunie" some distance off on the floe, but coming towards us. When he had approached within 600 yards his curiosity appeared to be satisfied, and he seemed inclined to show us his heels, perceiving which I landed on the floe, and, running as fast as possible, got him between the water and myself. His retreat being thus cut off, he jumped into the water, when, risking a long shot, I sent a bullet through his body. This apparently had little or no effect upon him, except to dye the water with his blood. Seeing the dingy coming up behind, he jumped on the ice, which gave me a better mark, and I fired again, striking him just abaft the fore shoulder, and rolling him completely over; he soon, however, picked himself up, and rushed towards me; but another bullet at about fifteen yards put an end to any evil designs he might have entertained towards me. Bear shooting alone on the floe is hardly a desirable or pleasant amusement, unless armed with a breech-loader. Whilst the others were employed flinching our game I walked some distance, and caught sight of another bear; but though I did my best to imitate a seal lying on the ice, I failed to allure my intended victim within range, and at last, suddenly taking fright, he turned round and scampered off.

We were just shoving off to return, when we heard the blast, and eventually saw the crown of a large whale. Double banking our oars,

ADVENTURE WITH A WHALE.

we pulled with all speed to the ship to make known the welcome intelligence, reaching her at about 4 P.M., and two boats were soon away in the direction we had indicated. Dinner was scarcely over before a fish was seen close to the ship, and all hands were called. He rose close alongside. Tom Webster pulled up—all was breathless excitement, everyone watching the result. He had a splendid chance, right over the fish, which we already regarded as our own. He fired, but, sad to relate, missed. The captain hailed from the nest to put in his hand harpoon; he stuck it in, but, fortunately as it turned out, had not time to bury it deep in the whale's blubber. Down went the fish with fearful rapidity, the lines fouled, and in another moment the boat would have been taken down. "Jump overboard for your lives!" shouted the captain from the crow's nest. At that instant the harpoon drew, and they were safe; but our fish was lost. Poor Tom! we were all more sorry for his misfortune in missing than for the actual loss of the whale. But we had no time to lament our ill-luck, for more fish were seen, and in less than half an hour we had the pleasure of hearing "a fall" cried, and of knowing that Jemmy Grey was fast. Seeing that it was a heavy fish, and likely to give trouble, Chester and I volunteered to go away in the dingy with the rocket-gun to kill it, an offer which was gladly accepted; Chester took the gun and I the steer oar, the doctor (Graham) and an old fireman, commonly called old Harry, forming our crew. We pulled close alongside the monster, which had by this time got three harpoons in its body. This seemed to make it wild; Chester fired. I swept the boat round, but the dingy, rather an unmanageable little boat in consequence of a very deep false keel, which had recently been put on, and also being a boat very ill adapted for the service on which we were employed, failed to get clear of the brute's tail, which it had thrown up out of the water on receiving the contents of our gun, and which, descending with terrific violence, just caught the gunwale of our boat, knocking me over the stern. Before coming to the surface I imagined the dingy had been smashed to pieces, which would have been rather a bad case for us, as the other boats were some way off, and, also, fast to the fish; and no loose boat being near us, and with the temperature of the water only a few degrees above freezing point, I don't think that I for one could have kept up long, accoutred as I was in a heavy monkey jacket and sea boots. However, on rising to the surface, I had the satisfaction of seeing the dingy a couple of boats' lengths off, and the doctor (who had taken to the water, imagining that the tail was coming right down upon us) and myself were soon hauled in, none the worse for our ducking. If the boat had been one foot nearer the fish she would most assuredly have been dashed to pieces and we should have all been killed before having time to jump overboard.

However, we ended in having our revenge on the monster, though it was an hour and a-half before we succeeded in despatching it. Having no boat to assist us, the whole duty of killing the whale fell to us. Chester (an old whaler) used the lance in a masterly manner. I was not sorry to return on board and get some dry clothing. The captain was much relieved at seeing us safe, for he had been very anxious after having seen the doctor and me in the water. We soon had our friend alongside, and by midnight his blubber was on board. Three or four more such fish will fill us. The number of whales seen here reminds me of Parry's visit to Regent's Inlet, and of his remarking on the numerous fish that were seen. He then predicted that at no very distant period it would undoubtedly become a rendezvous for our whalers, when the fishery in Baffin's Bay began to fail, as whales will always in course of time leave a place where they continue year after year to be molested. Parry gave a very exact description of Neill's

Harbour, thinking the day would come when our ships would find it of service.[1]

Wednesday, July 30th.—The morning broke wet and disagreeable, with rain continuing until noon, when the weather cleared up, and a strong breeze sprang up from the north-west; but in consequence of the amount of loose ice in the inlet, the water was comparatively smooth. Our only danger now is, being caught by the ice and beset until a change of wind. We have been cruising all day off and on Port Bowen, Parry's winter quarters in 1824, but I have been unable to get sights, as there is nothing but an icy horizon. I have employed myself to-day in tracing off the land as it should be on the chart, and skinning birds, which is anything but a pleasant process.

Friday, August 1st.—We are now cruising off and on a large floe, stretching nearly from Port Bowen to Batty Bay. The wind has been blowing very fresh for the last two days from the north-

---

[1] He says: "I have been thus particular in describing Neill's Harbour, because I am of opinion that at no very distant date whalers may find it of service." "Prince Regent's Inlet will probably become a rendezvous for our ships, as well on account of the numerous fish to be found there, as the facility with which any ship having once crossed Baffin's Bay is sure to reach it during the months of July and August. We saw nine or ten black whales the evening of our arrival in Neill's Harbour."—*Parry's Third Voyage.*

west, which is driving the ice up the inlet at a surprising pace. During the time we have been "hove to" off a floe some five or six miles in extent, it has been drifting to the southward faster than we have ourselves.

The captain has now a new idea in his head, which is to follow the ice of the inlet until we reach Fury Point, when we shall probably come to open water, to push up the Gulf of Boothia, making a running survey of all the unknown land on its eastern side, through the Fury and Hecla Straits, down Fox Channel or the Frozen Straits into Hudson's Bay, and thus home. I need not say how anxious I am that he should adopt this route, the discovery of which would be of the greatest advantage to whalers; and, from what we have lately seen, the whales are all heading in that direction, so that we might fish as we go along. At five this morning, a little south of Batty Bay, some whales were seen, and one small one caught, though all the boats were away. Whilst we were employed in flinching our fish, with the help of one or two of the "Polaris's," I got a cast with the deep-sea lead, but no soundings in two hundred fathoms, temperature 29·8; surface 33·0, air 36·4. The fish we obtained, though it had only run out three hundred fathoms of line, was covered with soft mud, and this was what induced me to get a cast.

In the evening we steamed through a stream of pack ice into a large open water along the land, the "Camperdown" following us. The way in which we bored through the ice elicited the admiration of the "Polaris's" men, Schumann, the engineer, saying that if they had come across such ice they would have made fast and remained until it opened He said if they had only had a man like Adams as their sailing master, he felt convinced they would have reached the North Pole. We passed a large iceberg to-day apparently aground. I am at a loss to conceive from whence it comes, as to my knowledge there are no discharging glaciers, nor, in fact, any others in this neighbourhood. It must, therefore, have drifted up from Lancaster Sound, or perhaps even from Baffin's Bay.

Saturday, August 2nd.—At six this morning we passed Fury Beach, at a distance of about three or four miles. I was very anxious to land, but the captain was as anxious to look for fish, so we stood on past Fury Point into Creswell Bay. With my glass I could plainly perceive a quantity of stores, which seem to consist of casks, spars, and a heap of what sailors would call gear. Two boats were seen hauled up on the beach, but one appeared a perfect wreck. We could also see a large anchor, but no signs whatever of the ship herself. The land seems to be of the same Silurian limestone as at Port Leopold. I am afraid we shall not proceed much further to the southward, as the captain reports ice closing in, so that our chance is small of going up the Gulf of Boothia. Certainly, as far as I can see, there is open water to the south, and the ice is at least four miles off the west coast of the inlet. I obtained good sights again to-day.

Midnight.—We have just returned on board from a long, wearisome, and unsuccessful chase after a whale. Shortly after entering Creswell Bay some fish were seen. All hands were called, and all the boats sent away, the captain observing with glee, as we shoved off, that we should be a "full ship" that evening. I went as boat steerer in Bannerman's boat, going with two others after a fish towards the head of the bay. For eight hours did we chase and dodge that whale, sometimes getting within one hundred feet of it, but never near enough to risk a shot. We must have pulled at least a distance of twenty miles, first to starboard, then to port, then ahead, and now astern. I verily believe the fish had seen us, and was purposely leading us astray. We chased it nearly up to the head of the bay, and then out again towards the ship, when it again turned and went back. The captain, seeing our crew were nearly done up, recalled us, and sent three fresh crews away in the boats after the same fish, but I do not believe they will ever get her. We were not sorry to return on board, as we were getting both hungry and tired. There appears to be a sort of inlet at the head of Creswell Bay, but whether formed by a river or not I am unable to make out. A low spit runs to the northward, and the high land trends away to the back of it.

Prince Regent's Inlet—The "Arctic" fishing at the edge of the floe.

Chapter XVII.

FURY BEACH.

S*UNDAY, August 3rd.*—The boats returned at four o'clock this morning, and were, as I predicted, unsuccessful, having given up the chase as hopeless. As they pulled towards the ship, the whale gave three or four unusually heavy, and, as it were, defiant and triumphant blasts. Fearing the ice was closing in, the ship's head was pointed to the northward, and shortly after breakfast we were again off Fury Beach. This time the ship was "hove to," and we all went on shore.

Fury Beach is classic ground in the annals of arctic adventure. During the stormy month of August, in the year 1825, it was the scene of much heavy work zealously performed, of an exercise of most skilful seamanship, and of the final loss of Her Majesty's ship "Fury." On the 1st of

August, 1825, the "Hecla," commanded by Sir Edward Parry, and the "Fury," under Captain Hoppner, after breaking out of their winter quarters at Port Bowen, were making their way up Prince Regent's Inlet. Among the officers of that expedition were four men, besides the commanders, who have won a permanent place in our naval records. James Ross was there, Horatio Austin, Edward Bird, and Crozier, the future second in command under Sir John Franklin. Parry had a fine staff, and all were full of hope. Suddenly the ice came bodily in, both ships were forced on shore, and no open water could be seen from the mast-head. The "Hecla" was hove off and warped to a floe in the offing; but the "Fury" leaked so heavily that four pumps were kept constantly going; and it became evident that, if she was to be rendered sea-worthy, she must first be hove down. Parry adopted a most ingenious contrivance to form a sort of basin or wet dock during the operation. Anchors were carried to the beach, having bower cables attached to them, and passing round the grounded masses of heavy ice. The cables were floated by two hand masts and empty casks being lashed to them, so as to make them receive the pressure of the ice a foot or two below the water. This heavy work was completed during the night of the 5th. For several following days all hands were employed landing provisions, spars, and boats, unrigging the "Fury," and landing sails and booms, coals and stores. It was a most animated scene, and every soul was fully employed. Casks were landed by a hawser secured to the mast-head, and set up to an anchor on the beach, the casks being hooked to a block traversing the hawser as a jackstay, and made to run down it. Nothing could exceed the spirit and alacrity of every individual. The officers of the "Fury" messed and slept in a tent on the beach, while the men were lodged on board the "Hecla." By the 18th the "Fury" was completely cleared, and they were in the act of heaving her down, when a storm came on; there was a heavy sea; the protecting ice was worn away, the cables slacked up, and the basin thus lost its protection. The "Fury" was once more driven up on the beach by masses of ice; and by the 21st an icy barrier, three or four miles in width, separated the "Hecla" from her doomed consort. On the 25th, Parry, with several officers, went in a boat to Fury Beach, and it was then decided that it would be impossible to make the "Fury" seaworthy, and that she must be abandoned. Her boats were hauled up clear of the ice, and she was left to her fate.

When Sir John Ross undertook his expedition

in the little "Victory," he relied a good deal on the great store of provisions that had thus been piled on the desolate shore of North Somerset. He landed at Fury Beach on the 12th of August, 1829, and found the mess-tent of the "Fury's" officers still standing, where James Ross, who was with his uncle, had rested after those days of heavy toil just four years before. The ship had been carried bodily off, probably ground to atoms, and floated away to add to the drift timber of these seas. Ross carried off an immense quantity of stores, ten tons of coal, and sails for housings, and made sail on the 14th.

The Rosses passed three long years on the shores of Boothia to the south, during which time James Ross discovered the magnetic pole. At last the crew of the "Victory" had to abandon her, and retreat to the north in order to seek safety by reaching the whalers in Baffin's Bay. The friendly stores of Fury Beach were to be their half-way resting place, and they arrived there, travelling wearily over the ice, on June 30th, 1832. Here they built a house, with two rooms, which was named "Somerset House," and then set out in three of the "Fury's" boats for the northward. But they were stopped by the ice near Port Leopold, and resolved to return to Fury Beach, and there pass the winter. The boats were left in Batty Bay, and the Rosses went back on foot.

establishing themselves, with their men, in Somerset House, on the 7th of October. Here they lived on the "Fury's" stores, and only lost one man during the winter. In July, 1833, they finally departed, and were picked up by a whaler in Lancaster Sound.

In the spring of 1849, Lieutenant Robinson, of H. M. S. "Enterprise," then wintering at Port Leopold, undertook a journey southward, along the coast of North Somerset, and reached Fury Beach on June 10th, in the hope and expectation of finding traces of Sir John Franklin's people; for Crozier, the captain of the "Terror," had been a midshipman in the "Hecla" with Parry in 1825, and had shared in the toil of landing all the stores. It was fully anticipated that at least some of the companions of Franklin would have made for this spot, but there was not a trace of them.

The private searching vessel, "Prince Albert," under Mr. Kennedy, and with Lieutenant Bellot of the French navy on board, wintered during 1851-52 in Batty Bay, on the west coast of North Somerset; and in January, 1852, Kennedy and Bellot undertook a sledge journey to Fury Beach to lay out a depôt. They reached the spot on the 8th, and found the framework of Somerset House still standing and entire, but the covering was blown to rags. They lighted a fire in the old stove, supped and dozed over it, and then set out on their return journey to the "Prince Albert" in Batty Bay. In February, Kennedy and Bellot started on a more extended journey, and this time they remained at Fury Beach from the 5th to the 29th of March, taking some of the stores, which they found to be in excellent preservation. They then set out again, discovered Bellot's Strait, and returned to Batty Bay, after having travelled right round North Somerset.

When Sir Leopold McClintock went down Prince Regent's Inlet in the "Fox," he was off Fury Beach on the 20th of August, 1858, but did not land. In March, 1859, however, Mr. Allen Young left the "Fox" at Port Kennedy with a party consisting of the cook, two Esquimaux, and some dogs, in order to obtain provisions at Fury Beach. He reached the place on the 22nd, when the ground was covered with snow, so that it was necessary to dig down to the casks. After much trouble he obtained what was required, about 8 cwt. of sugar, and some preserved vegetables. His report to Sir Leopold McClintock was, that the stores were so covered with deep snow that it was impossible to take an account of them, but that he had found some casks of sugar, tobacco, peas, and flour, a very little coal, and two boats —one a four-oared gig, the other a cutter with the side cut out as if for making a flat sledge. Allen Young returned to the "Fox" on the 28th, after a very severe journey across heavy hummocks in Creswell Bay, where his sledge broke down, and he became snow blind, and had to be left alone in a bag in the middle of the bay for forty-eight hours, whilst his companions made progress with the heavy load.

No living soul had landed on Fury Beach since March, 1859. I looked upon it as classic ground, for here Parry, Hoppner, James Ross, Bird, Austin, and Crozier had displayed all the finest qualities of seamen in 1824. Here the Rosses wintered in 1833; here Kennedy and the gallant Bellot had rested in 1852; and this was the scene of Allen Young's severe work in 1859.

Casks, spars, rigging, and a perfect assortment of ship's stores were strewn about in all directions. Two boats were hauled up on the beach. One was in pieces, as described by Allen Young; but the gig, on a pinch, might be made seaworthy. It is a 22 foot four-oared gig, painted black with a green ribbon, the paint appearing tolerably fresh. The other had been a larger boat, about 30 feet long. It has "W 1824 ↑" on its stem—being a Woolwich-built boat. On the stern the name of *G. Fowler, June 10th*, 1849, was cut: a marine belonging to the party of Lieutenant Robinson, under which our doctor carved his own name: *A. Graham, August 3rd*, 1873. The gig had been built either at Devonport or Deptford, and had on its stem "XXII. ↑ D. 24." The remains of Somerset House, which had fallen or been blown down, were lying between the two boats, the framework being composed of spars. The running rigging of the "Fury" had all been unrove, and was lying on the beach, jagged up in five fathom lengths. The topsail yards were also on the beach; but there were no signs whatever of the hull of the ship, or of her lower masts and yards. She must have been crushed up or carried away by the ice many years ago. Three bower anchors were on the beach, one of them having a large 10 inch hawser attached to it. They were 22 and 25 cwt. anchors, the latter being a patent one, marked *R. F Hawkins.* There were many hundred tins of preserved beef and vegetables, and also what one of the men informed me was "*consecrated*" gravy, all in an admirable state of preservation, after a lapse of nearly fifty years. The flour was all perished, but the sugar and tobacco appeared to be good. One would have imagined it had been the wreck of a whaler instead of a man-of-war, for a perfect set of whaling implements, including harpoons and lances, was on the beach. Readers of Parry's voyage will remember that Ross and Sherer succeeded in killing a "payable" whale off Port Bowen.[1] The

---

[1] Parry's "Third Voyage," p. 89.

metal powder-cases had all been broken open, and some of the powder was scattered about. I tried some of it, which ignited, but it burnt slowly.

On landing I found a pair of large deer antlers, of which I made prize. As I could not find the skull, they had probably been shed there by the animal itself. The marks of bears' teeth and claws were plainly visible on some of the casks, the wood of one containing flour having been literally gnawed through.

Seeing a cairn near the water's edge, I hurried towards it, and quickly demolished the heap in the expectation of finding some record, but, after an hour's hard work with pick and shovel, I was horrified to find that it was a grave, the body having been sewn up in canvas instead of a coffin. I carefully replaced everything, endeavouring to give the heap more the shape and appearance of a grave than a cairn. It must have been the body of Chimham Thomas, the carpenter of the "Victory," with Sir John Ross, who died on February 22nd, 1833. His is the only body that is buried on Fury Beach. He was aged forty-eight, and his constitution had been undermined by long service in the first Burmese war, and on the American lakes, before he ever made an arctic voyage. It is very strange that this poor man should have been buried so near the water, for a more desirable site could easily have been found further inland.

Two 32-pounder carronades and a small gun-carriage, with a large amount of shot of all descriptions, were lying about. A pistol was also picked up, one of the old navy pattern. I found an old rusty knife, which, with a good harpoon, a broken pair of binoculars, left by Lieut. Robinson in 1849, and the antlers, will be my souvenirs of Fury Beach.

Leaving this curious scene, I scaled the hills in rear of the beach, accompanied by Dr. Bessels. They rise in terraces to a height of about 500 feet, stretching away landward in long and smooth undulations. From the summit I obtained a view of the opposite shore, and could distinctly see Cape Kater and Cape Garry. The ice seemed to remain in the centre of the strait, leaving a clear open water along the coast. We found several granitic boulders on the top, conclusive evidence of the upheaval of the land, which is composed of Silurian limestone. I was fortunate enough to pick up a few fossils, and brought away several geological and botanical specimens.

We returned to the ship at three o'clock, and she stretched away towards the ice in the centre of the strait. It turned out a very wet afternoon and evening, rain falling heavily at times. The sun is gradually declining, and it will not be long before we lose our midnight sun, which will be a source of regret, as I do the greater part of my work at night, when everything is quiet. I was much amused to-day at the curious fancies which sailors sometimes take into their heads. One of the men was suffering from a severe cold and sore throat, and, having given the doctor's prescriptions a trial without alleviating his symptoms, he determined to try a remedy of his own. Accordingly, having obtained a raw salt herring from the steward and taken the bone out, he applied it to his neck, tying a handkerchief over it, and keeping it on all night. He assured me in the morning that he had derived much benefit from its effects.

King Ducks.

Chapter XVIII.

CAPE GARRY.

M*ONDAY, August 4th.*—Last night, just as I was going to bed, Bannerman, whose middle watch it was, came down and asked me if I would like to go away with him after *unies*, two having already been harpooned, adding that a bear had been seen on the ice. I did not require much persuasion to go, and taking my rifle, we were soon pulling about after narwhals, Bannerman as steersman and I as harpooneer. Soon we espied a large bear, apparently asleep on the floe; but the ice being loosely packed, we were unable to approach within 500 yards with the boat. We therefore jumped out on the ice, and endeavoured to make our way towards the animal. This was by no means easy, as the floes were in some places so far apart as to render it impossible to jump

ADVENTURE WITH A BEAR.

Page 254.

across; we were, therefore, compelled in these places to push the floes over with boat-hooks, which we had taken the precaution to bring with us. Great care had also to be taken to avoid falling through, as the ice was very thin and treacherous, added to which it was drifting rapidly to the southward; so, had either of us fallen in, the ice would have passed over before we could have had a chance of getting out. After some little time we got within 150 yards of "Master Brunie," who had been lying quiet all the time. Having arrived at that distance, in our anxiety to get near, we jumped on a piece of rotten ice, which instantly gave way with an unpleasant crumbling noise. Fortunately, we were able to scramble out, wet only to the waist. The noise, however, disturbed Bruin, who, raising his head, surveyed us intently. Bannerman fired, but missed, which caused the bear to get up and meditate a retreat, when I fired, striking him just behind the head, and rolling him completely over. He gathered himself up again pretty smartly, when I again fired, the ball passing through his neck. By this time, having run on in a frantic manner, splashing through water and ice alike, I had come close up to our friend, who, seeing me, rushed, open-mouthed, towards me. When he got within five yards' distance I fired, the bullet striking between the two eyes, and at once terminating his sufferings. He is the largest we have yet shot—fully ten feet in length. After flinching him we had the unpleasant task of dragging the skin down to the boat. If our journey towards the bear was bad, our return was far worse. On arriving at the boat we found, to our chagrin, that she was completely beset, the ice having closed in all round. To wait longer would only make matters worse, so we had to look our difficulties resolutely in the face and commence action. For upwards of an hour were we breaking through thin ice, or hauling the boat bodily up on a large floe, dragging her across, and launching her on the other side. It had one good effect, which was that of keeping us warm, so as to counteract the effects of our wetting. Matters were beginning to look rather serious, steady rain and thick weather having set in, when, fortunately, they saw from the ship the unpleasant predicament in which we were placed, and the captain, ordering steam to be got up, bored a passage to us through the ice, and so relieved us from our troubles. It was past 3 A.M. when we got on board. We were in a most filthy condition, for the boat we were in was the "mollie" boat, that is, the one employed alongside a whale during the process of flinching, and everything was covered with a thick coat of dirty grease. It has been a disagreeable, thick, rainy day, but I am glad to say we are again steaming to the southward, and are this evening in Creswell Bay ; and though there seems no chance of going north, yet much may be done in Prince Regent's Inlet.

Tuesday, August 5th.—This morning found us well inside Creswell Bay, and, being clear, I succeeded in getting good sights and bearings, but, unfortunately, the horizon was rather hazy. Cape Garry appeared to me to be placed too far to the southward and eastward. It was originally laid down by Parry, in 1819, nearly in its true position, although he only took a bearing from some distance. Its removal further south appears to have been due to Sir John Ross, and is certainly erroneous. I was in great hopes of landing, and had everything ready to fix its position, but a reef appearing to run off the point and the water shoaling, warned the captain of hidden dangers, and he therefore, much to my disappointment, though perhaps very wisely, tacked and stood to the northward. However, it would have been foolish to have acted otherwise, and to risk a full ship close to what was considered a reef. We got soundings about six miles from Cape Garry in 30 fathoms, hard bottom, no particle of any description adhering to the arming of the lead. A boat was sent away to sound in the direction of the cape, obtaining 25 fathoms, gradually shoaling to 4 within three or four miles of the land, which projects out from the cape about two or three miles, forming a long, low spit. The bottom appeared to be sandy, a minute portion being brought on board that had been drawn up with the lead. I examined it through a magnifying glass, and found that it was evidently formed from sandstone, though I detected a small fragment of shell. It seems to me rather inexplicable that the bottom should consist of sandstone, whilst the land is composed of limestone. The bottom I obtained at Elwyn Inlet, at a depth of 190 fathoms, was limestone the same as the land. The ice appears not only to be drifting to the southward, but also to be breaking up, so I trust we may yet have a chance of pushing up the inlet. I should have liked to have obtained more soundings to-day, but the ship was going too fast through the water.

Thursday, August 7th.—We passed Fury Beach at midnight last night, and I doubt very much whether we shall go to the southward of it again. The captain kept us up and amused us till past two this morning, reciting Shakespeare, singing songs, performing the showman, and indulging in various other accomplishments too numerous to mention. He has certainly a wonderful memory. In the afternoon, several of what the captain calls "sword fish" were seen, and the boats were sent to attempt the capture of one; but they were flying so fast through the water, and were so wild, that our boats were unable to approach within striking distance. These fish appear to be the regular grampus, having a dorsal fin, and being apparently about twenty feet long. They are the whales' most inveterate enemies, scaring them away directly they appear in their neighbourhood; and the captain is convinced that in consequence of our meeting them we shall see no whales for some time. Their blast is much heavier than that of a narwhal. No one on board this ship has ever seen one of these sword fish caught. The ice both yesterday and to-day has been drifting rapidly to the northward, that is out of the inlet, and is also breaking up quickly, and in a few days, if we remain, I expect open water will be visible as far as it is possible for us to see up the inlet.

Friday, August 8th.—A miserable, cloudy, wet day. I almost wish winter would set in, as rain on board these ships is most disagreeable—far worse than snow. Shoals of grampus were seen to-day, which is, I am afraid, a strong indication that we shall get no whales for some time. Our worthy skipper is somewhat changeable in his plans and ideas. Yesterday he was all for going north and visiting the "Polaris;" to-day he is going to remain in Regent's Inlet until the end of this month, and push up to the head of the Gulf of Boothia as far as Committee Bay. Whichever he does will afford me great pleasure, though I should prefer the former. This thick weather is much against taking observations. I must hope for a change for the better when we get upon unexplored ground.

Sunday, August 10th, 3 A. M.—I have just returned on board from a long, cold, wearisome pull after fish. We have been again amongst them, but though successful in one instance, this (yesterday) has been a most unlucky day. We passed Fury Point early in the morning, and I obtained excellent sights, fully corroborating all my previous work. Cape Garry is really fifteen miles to the northward of the position assigned to it on the chart, and very nearly in the position originally given to it by Parry. The ice is much further to the southward and westward than it was a few days ago. At noon we were abreast of Cape Garry, about twelve miles distant. The land was plainly visible to the southward, trending in a S.S.W direction, and beyond that low land was seen, which I take to be the north coast of Boothia Felix, so that we have really seen the continent of America. At 1 P. M., as I was busy working out my sights, a great commotion overhead told me that whales were in sight, and on going on deck the captain hailed from the nest to say he saw any amount, and was certain that we could get enough to fill us up. The ship was hove to, and all boats lowered away, I steering Bannerman. We had not left the ship five minutes before a fish rose close to us. I swept round; we pulled up quickly; the mate fired, and we were fast. About three minutes afterwards Harky Hunter got fast, and then Jemmy Gray. Three fish in less than ten minutes! Unfortunately they all took to the ice, and it was a work of great difficulty, not unattended with danger, to get near them. Eventually one was killed, but the other two got away. Ours took out nearly two miles of line, which must have chafed on the rocky bottom, as it parted about fifty fathoms from the harpoon, so we had the delightful occupation of hauling in the line with nothing at the end of it. We got back to the ship about eight, and after refreshment started again on an unsuccessful expedition after whales. At midnight a fresh breeze sprang up ahead as we were returning, which made it bitterly cold, and also unpleasant to steer, the sea rising and nearly unshipping the steer-oar at every toss, and we came on board wet, cold, and discontented. I have now just got up, and find that we are off Cape Garry, and the day fine and clear. The captain made me supremely happy to-day by consenting to my proposal of taking the dingy, with the doctor and a couple of volunteers, for the purpose of going up Bellot's Straits, and to be away three or four days. I have no doubt in that time we might do a good deal. I picked out a piece of gneiss to-day from the gun-harpoon that Jemmy Gray fired yesterday, and which had been dragging along the bottom.

Monday-night, August 11th.—Just returned from a short but most successful exploring expedition, very tired, very sleepy, and all my bones aching. Last night, being about eight miles off Cape Garry, the captain consented to my taking the dingy to go on shore, telling me to look out for a red ensign being hoisted at the fore, as a signal for my immediate return, which he would only hoist if he saw the ice setting in. Accompanied by Dr. Bessels and Hermann (one of the "Polaris" men, a German) and taking sufficient provisions to last a week, we left the ship at 8 P. M. and stood in with a fair fresh breeze, intending to land in Fearnall Bay, to the southward of Cape Garry. When we arrived within a mile of the shore, I found the water so shallow as to utterly preclude the possibility of landing, the dingy nearly grounding. Bearing up, and running along the coast to the southward, we managed to land round a point about six miles from Cape Garry.

Leaving Hermann in charge of the boat, the doctor and myself started to explore. Walking was by no means easy, in consequence of the numerous streams we had to cross, which, running down from the hills, emptied themselves into the sea. Having long boots on, this difficulty was overcome by my taking the doctor on my back and carrying him over Unfortunately one river was deeper than I imagined, and the water coming over my knees, made me wet for the remainder of the day Having walked rather more than two miles, I saw with my glasses what I imagined to be the carcase of a whale washed up on the beach, but what was my surprise, when we arrived at the spot, to find traces of a large Esquimaux village, most of the huts having been actually composed of the ribs and trunks of whales. We counted no less than thirty-four huts, seven of which had originally been made of stone, seven very old ones, and the remainder built from the bones of whales. Sixty skulls had been used to form the foundation and entrances to the "igloos." It would be very interesting to know how these remains of whales had reached this place, whether they had been washed on shore, or whether they had been killed by the inhabitants of the settlement. Comparing these with the bones which we found at Fury Beach, and which we knew to be about fifty years of age, I should say that no Esquimaux had been to this locality, at any rate to reside, for fully eighty or a hundred years. I picked up a couple of pieces of bone, which had evidently been used either for a sledge or a kyak. We observed traces of deer, bears, foxes, and lemmings, and saw plenty of ducks and brent geese. About 1 A. M. we returned to the dingy, and having refreshed ourselves with Australian mutton and biscuits, made sail again to the southward, the ship being now nearly out of sight on the horizon Having sailed about six miles along the coast, we landed on a point which formed the north extreme of a large deep bay, unmarked on the chart; where I put up my horizon and got my first set of sights at about 3 A. M. The tide falling rapidly, almost before we were aware of it, left our dingy high and dry, so we were compelled to wait for the flood tide. The doctor and myself started for a walk towards some high hills in the interior. Hoping to see some big game, I took my rifle with me. Having walked a little over six miles, and being rewarded by finding limestone abounding with fossils, and other geological specimens, I saw four deer, a buck, two does, and a fawn, about 300 yards down a gentle incline between two hills. Approaching as stealthily as possible, I managed to get within 150 yards, and fired at the buck, my bullet taking effect, though not sufficiently to stop him, and I had to fire more than once before I brought him to the ground, the others making off. Now came the question, what were we to do with it? The idea that two men, not over fresh, should drag a stag weighing 200 lbs. over very rough ground, a distance of more than six miles, was simply absurd; so we resolved to return to the boat, and bring her round to a bay nearer by half a mile than the place where we had left her, and then the three of us start and carry our prize down. Accordingly, acting on this decision, we cleaned it with our penknives, and dragging it about a quarter of a mile to the summit of a hill, we left it in a conspicuous place and returned to the dingy. To ensure our finding it, I tied my pocket-handkerchief to one of its antlers, which blew out famously in the breeze. It was eleven before we reached the dingy, very tired and very hungry, and regaled ourselves on sardines and Australian meat, washed down with a delicious glass of beer. Unfortunately it had clouded over, and I was only able to get one sight, and that none of the best. After a couple of hours rest, we managed to get the boat afloat, and sailed a little further up the bay, when making her fast, we all started to bring our game in. We had walked barely half a mile, when I saw a fine buck jumping and skipping about not very far from us. Directing my companions to halt, I approached warily; but the deer, not liking my looks, began to scamper off, when I took a chance shot, the bullet falling short, but splashing the water alongside him, for it was marshy ground. This had the effect of stopping him, and he turned round to look from whence it came, offering me a fine shot. Raising my sight to 300 yards, I fired; the bullet passing through his head and killing him at once. He proved to be a fine fat buck, and; overjoyed at our success, we slung him on an oar and carried him down to the dingy, mighty glad to get there, as my shoulders were getting very sore. We then started a second time in search of our first deer, and eventually found it about five miles and a-half from the boat. By this time we were all pretty well done up, so we determined, as the easiest way of carrying it down, to cut it up, which was accordingly done. Hermann took the hind quarters, the doctor one fore quarter, I the other and the head. This, with my rifle and ammunition, made a tolerably heavy load for a weary man. I don't believe I could have gone a mile further, and was most thankful to get down to the boat. I intended going on to the south point of the bay, and then on to Bellot

Straits; but seeing the red ensign flying on board the ship, we were forced to return, arriving at 7 P. M., after a most interesting and successful twenty-four hours' cruise.

Deer's head.

Chapter XIX.

HOMEWARD BOUND.

T*UESDAY, August* 12*th.*—When we came on board last night we were received with the very pleasing information that a fine whale, which would probably yield seventeen tons, had been caught during our absence. The men have been engaged since four this morning in making off the blubber from the last fish. At about nine a tremendous cheering announced the fact of the ship being full between decks, and nearly ten tons over, which will be stowed in some small supernumerary tanks with which the ship is supplied. We are to all intents and purposes a *full ship*, and have on board the largest cargo of oil that has ever gone home from Baffin's Bay, so it will not be long before we are homeward bound. All day I have been employed skinning the birds I was fortunate enough to obtain yesterday, which consist of a king duck and duckling (*Somateria spectabilis*), a long-tailed duck (*Fuligula glacialis*), and an arctic tern (*Sterna arctica*), all good specimens.

I shall be able to make a fair delineation of the trend of the land south of Cape Garry, with correct latitude, which I have again confirmed to-day by sights when abreast of the cape. I do not understand its position being so much out on the Admiralty chart. I feel uncommonly stiff to-day, and all my bones are aching from carrying the deer. We must have walked yesterday upwards of thirty miles, which is a good stretch for one who has been cooped up on board ship for so long as I have.

HOMEWARD BOUND.—I hardly know whether it gives me pleasure or not to write the above words. I am certainly delighted at the idea of going home, but wish I could have done and seen more before returning.

At five o'clock this afternoon the captain came into my cabin, asking if I could spare him a few minutes' conversation, during which he told me that he hardly liked risking the ship any longer amongst ice, as she was very deep in the water, and would, consequently, strike very heavy; that he had on board the best cargo that had ever left this part of the world, and that he had only a few small tanks to fill, every other, even the bread and water tank between decks, being full; and that he had almost made up his mind to go home. Of course I told him he must judge for himself, though, as far as I was concerned, I should like to go north. He replied, that he was most anxious to oblige me, and would be guided by circumstances. If there was a southerly wind outside Lancaster Sound he would try and go up. Having made up his mind to go home, he sent for all the harpooneers, saying to them as they came aft, "What do you say, boys, home, or another fish?" They all with one voice cried, "Home!" and gave three cheers, which was soon taken up by the men, and continued for some little time, everyone being in the highest spirits. Steam was raised, and the order given to go ahead; and now the fan is making the ship throb again as we steam past Fury Point and down Prince Regent's Inlet. All my hopes are now centred in a southerly wind, and plenty of it. We have lost our midnight sun, and for the first time since crossing the arctic circle I am writing by the light of a candle, though it is quite light enough to do anything else but write.

Wednesday, August 13th.—Unmistakable signs of approaching winter were observed last night, the ship having to steam through a large extent of bay ice.[1] It is quite surprising the quantity of pack ice that is now in the inlet, much more than when we came up. The late northerly winds must have set it in from Barrow's Straits. We attempted to shape our course from Fury Point to Cape York, but the ice proved such an obstruction, that we had to steam up along the west coast until almost off Port Leopold, before we could strike across the inlet. At eleven this morning two bears were seen on the ice ahead, one very large and the other small. On observing the ship, they took to the water, swimming in different directions. The ship was stopped and the boats lowered, I in the one that went in pursuit of the largest. We had a great deal of trouble in getting near our victim, having to haul the boat through and over large pieces of ice. Eventually we came up pretty close. My companion, who had the first shot, missed; I then fired, killing the bear instantaneously. He proved a perfect monster, the largest that anyone on board this ship had ever seen, measuring over 10 feet, and weighing about 700 lbs.

The ship is beginning to assume quite a different appearance, all the coal and whalebone having disappeared off the quarter deck, though

[1] Bay ice is newly formed ice.

it is still lumbered up with casks. No attempts at cleaning have yet been made, beyond scraping the ship—the deck, sides, paint work, skylight, and everything—long layers of greasy substance coming off with each scrape. It will be a great novelty to see the decks washed. The last time such an operation was performed was at Upernivik, more than two months ago. The day has been very hazy and cloudy, which entirely prevents my completing my observations along this coast. We are obliged to keep clear of ice, and a long distance from the shore. I much fear nothing in the way of exploring will now be done by this ship, and that we shall go straight down Baffin s Bay, not even going through Navy Board Inlet, which the captain always said he should do, if it were only for the sake of getting some salmon. However, he has been asleep all day, so I have heard nothing further of his intentions. At 7.30 this evening we passed Cape York, and shall probably be off Cape Hay to-morrow at noon, when the captain must determine what he is going to do. Half an hour ago two bears, a mother and cub, were seen on the ice, and seizing my rifle, I jumped into one of the two boats that were ordered to be lowered to effect their capture. On pulling up to them, the old bear made a savage attack on the boat, and would very soon have got in and cleared it had I not put a bullet through her head, killing her instantaneously. Taking the old one in tow, we proceeded to secure the little one alive, and eventually hoisted him on board, though not without some trouble, as it was of a most pugnacious disposition, snapping at everything that

Cape Hay.

came in its way, and roaring and bellowing like a young bull. It is a fine little cub, about two months old. The captain intends taking it home alive.

Thursday, August 10.—At ten this morning we passed the mouth of Navy Board Inlet, the captain intending to steam out until he falls in with the ice, when he hopes to pick up another fish, which will completely fill us. At noon passed Cape Hay, and at four we were off Cape Liverpool, when the wind that had been gradually freshening since the morning, was so strong from the southward, that we were obliged to stop steaming, and put the ship under reduced canvas, a nasty sea getting up at the same time. Barometer low, and falling. In the afternoon we bent the mizen-topsail, and unrove the cant and spek tackles. It is very heart-rending getting such a strong head wind, especially at the commencement of our homeward journey. The poor little bear seems to be in a very miserable state, but whether it is in consequence of its captivity, sea-sickness, or pining for its mother, it is difficult to say.

Saturday, August 16th.—Yesterday was the most trying day I have spent on board the "Arctic."

The wind had gradually increased to a strong gale, accompanied by a heavy sea. As we had come out without making the necessary preparations for bad weather, we were in a pretty pickle. I had not long been in bed when, about 2 A. M., the captain burst into my cabin, and with his usual, "Are you waking, captain?" informed me that the dingy had been washed away, and one of the whale boats badly stove. Going on deck, I found things were not quite so bad as represented. The little dingy on being hoisted in-board, was found only slightly damaged; the barometer was very low, lower than we have experienced this cruise; it was blowing hard from the southward, and the ship labouring heavily under reefed topsails and foresail. Between decks everything was in a sad state: tanks, whalebone, and seamen's chests were lying about in fearful confusion, and it was with no little difficulty that the things could be properly secured. To-day affairs are looking a little brighter; both wind and sea have gone down considerably, though the barometer still remains unusually low. At 2 P.M. we found ourselves exactly in the same position we were forty-eight hours ago, namely, off Cape Liverpool. We shall be a long time getting home at this rate. Perhaps the low barometer indicates a change of the wind.

Sunday, August 17th.—The wind gradually subsided until it fell altogether towards the evening, and now we are, as we suppose, fifty miles off land, about abreast of Pond's Inlet, steaming toward the south. Snow and rain fell at intervals during the day, which was thick and cloudy; we steamed through several extensive streams of loosely-packed ice, gaining once more smooth water, which is a great comfort. Poor little Bruin is not at all reconciled to his captivity, and has been vainly endeavouring to make a hole through his cask with both teeth and claws. He is a savage little brute, seizing anything within his reach, and is wonderfully strong and quick for so small an animal.

Monday, August 18th.—Several of our old friends, the icebergs, are now in sight in various directions, otherwise there is not a vestige of ice to be seen; no pack nor stream-ice anywhere. The captain has never seen Baffin's Bay so clear, and this augurs well for a good open season next year, as it causes a free passage for the ice to come out of Smith's Sound, while, in case of southerly winds, the ice to the northward will more readily be broken up. Everything points to the ensuing year as being *the* season for Arctic exploration. With a stout ship, and a well-organized and an efficiently conducted expedition, there is no reason why that hitherto unapproachable spot, the North Pole, should not be reached in a couple of seasons. This morning we sighted the land about Coutts Inlet, but at such a distance as scarcely to be able to distinguish anything. It appears high, undulating land, and is thickly coated with snow. The rapid manner in which changes take place in these regions is most remarkable. This morning, with the exception of a few bergs, there was no ice visible, but this evening we are amidst heavy pack-ice, some of the floes being of great magnitude, which is causing us no little trouble to get through. It has evidently been set up by the late strong southerly winds. We are about thirty miles off Cape Adair. It is most annoying being such a distance from land, for, without making our passage longer, it would have been so easy to

Bear and White Whale.

have steamed along the coast, and we could then have taken some useful observations. This comes of trusting to a compass, when the variation amounts to 130° I had another chase after a bear this afternoon, which we suddenly disturbed in the middle of a feast. He had been so scared by the sight of the ship, that he went away over the ice and through the water at railway speed; and though we pursued him for half an hour in our boat, failed to get within shot. We revenged ourselves by taking possession of his dinner, which was found to be a white whale about fifteen feet long. We removed the blubber, the most luxurious part of his repast. This whale must have been hauled up on the ice by the bear. The strength of these animals is truly astonishing.

Shortly after tea a walrus was seen on a piece of ice, and I went away as boat-steerer in the boat that was lowered for the purpose of effecting its capture. It was a huge monster, weighing at least a ton. We sculled down quietly upon it. When just within shot the creature lifted its head to take a look at us, which movement broke the piece of ice on which it lay; it was gradually disappearing when Deuchars fired, but unfortunately missed, the harpoon striking over. I was sorry we were so unsuccessful, as it was a gigantic brute, with tusks over two feet in length.

Tuesday, August 19th.—Thick and foggy the greater part of the day, the ship steaming and sailing through extensive fields of pack-ice, and along the edge of very heavy floes. Both yesterday and to-day we passed an immense number of grounded icebergs, some of great magnitude, and this will probably account for the presence of so much ice. The floes have been prevented from drifting out by these huge mountains, which most effectually bar their progress. At about 4 P.M. we sighted and communicated with the little "Victor," and took on board her home letters. We also saw the only foreign whaler out, the "Harold," a Norwegian. She is clean. At 8 o'clock we saw and communicated with the "Tay," and from Capt. Gregg's account of the state of the ice I much fear we shall have great trouble in getting into the east water. He says it is impossible to get to it by going south; unfortunately, we have a northerly wind, and therefore must take advantage of it; I have no doubt we shall be able to push through in about the latitude of Home Bay or Exeter Sound.

We are now nearly in the latitude of Cape Kater of the whalers (Cape Raper of the chart), but nearly thirty miles from land. We see several of the other ships, but it has come on so thick that we must postpone communication with them until morning. It is now midnight and nearly dark.

Wednesday, August 20th.—The morning broke fine and clear, and found us surrounded by nearly all the whaling fleet, two or three of which were chasing fish. We have now taken on board from the "Ravenscraig" all the rest of the crew of the "Polaris," excepting three that the "Intrepid" has on board. With the exception of Morton, all the men we received on board to-day are grey-headed, and certainly not the sort that one would pick out to go to the North Pole;—I should say the average age of the four would be over fifty years. We now number thirteen in the cabin; seven sit down at the first table, and the others when they have finished.

We are threading our way through miles of loose pack-ice towards the southward. It is a dead calm, and we are compelled to have recourse to our unfailing kettle. I hope the coals will last out. This evening at 7.30 I witnessed for the first time a most perfect and brilliant parhelion, the angle between the true and mock suns being 22° 17′ 30″ and the true altitude of the former 5° 24′ 30″; the mock sun was to the southward of the true one. It is a belief amongst old Arctic sailors that these phenomena always precede bad weather.

Cape Warrender.

CHAPTER XX.

CONCLUSION.

THURSDAY, *August 21st.*—We are all, especially myself, sadly disappointed with the progress the ship has made during the last twenty-four hours. The uncertainty of everything in these regions sorely tries one's temper. Last night we got into a fine open water, which we were all convinced would lead us out into the east water somewhere abreast of Home Bay; and congratulated ourselves accordingly. But this afternoon disclosed, to our great disappointment, ice, heavy solid floes, stretching away from the land to the eastward as far as the eye could see, without even a crack or anything approaching a lead through. We are in a regular *cul de sac*. To add to our misfortune, a fresh breeze has sprung up from the eastward, which will pack all the ice tight up against the land floe, and we may consider ourselves lucky if we avoid getting beset and being jammed up here for a few days. Our only way of escape is by retracing our steps to the northward, and getting round the north end of the middle ice; but the weather has become so thick, accompanied by snow, that it is impossible to see our way out, and we are now lying, hove to, waiting for it to clear. We received a visit to-day from Captain Gravill, of the "Camperdown," bringing us his home letters. At noon to-day we were abreast of Cape Bisson, and I obtained tolerably good sights.

Friday, August 22nd.—This detention is most annoying. Ten days ago we considered ourselves homeward bound, and, anticipating a quick run across the Atlantic, were looking forward to being in Dundee by the 1st of September; instead of which we are now jammed in a water hole off Cape Kater, with at present no prospect of getting out for some time. The spirits of all on board are in consequence rather depressed. Snow has been falling heavily all day. During the afternoon the wind gradually went down, and we have now a light northerly breeze, and are seeking a passage out through the ice to the eastward.

Sunday, August 24th.—Retracing our steps to the northward. At noon we were as far south as Cape Bisson, but from the masthead nothing appeared but an impenetrable sea of ice, a second Melville Bay, through which there was no chance of forcing a passage. The helm was accordingly put up, yards squared, and we are running north before a light southerly wind. I am afraid it will take us at least two days to get round the north end of the middle ice.

Wednesday, August 27th.—The last three days have added nothing to our progress. In fact, our prospects of getting home are not nearly so bright as they were fifteen days ago, when the announcement that we were homeward bound was received with three cheers by all hands. We have unfortunately seen *fish*, and our captain seems bent on getting more; he says, "he will be happy with ane or twa mair." The weather is thick, and raining heavily; we are about seventy miles from land, somewhere in the latitude of Cape Hewitt, and are made fast to a floe, taking in water. It is next to impossible to do anything. Euchre commenced at half-past 9 this forenoon, and even with my cabin door shut I cannot avoid hearing incessantly from the adjoining cabin, "What's trumphs?" "Steward, a mouthful more of that brandy;" "Who played the rag?" "That's nines;" "Guess you're considerably euchred;" and various other expressions in Scotch, Yankee, and German accents. Occasionally this is varied by a song or story from the skipper. Oh! for a fresh southerly wind, with no fish to be seen, and we might have a chance of getting round the middle ice into the east water, and then bear away for home.

Thursday, August 28th.—Still thick, foggy weather. We have been for the last twenty-four hours threading our way through loose but heavy floe ice towards the north, making a little easting. If we do not see any fish, I really believe we shall soon get either into the north or the east water, and then there will be no further cause for delay, and every day will lessen the distance between ourselves and Dundee. The ship strikes the ice very heavily, and it is impossible to avoid occasional collisions, which make the old vessel stagger and reel, shaking everything on the table. Whilst lying in bed at night, after one of these shocks, one can distinctly hear the ice cracking and breaking up, and scraping along the ship's side as we forge our way through.

Saturday, August 30th.—The sun favoured us with his presence for a short time to-day, of which I took advantage to get some sights. We are in about the same latitude as Cape Adair and Black Hook, but not so far to the east as we had hoped. There is now plenty of water, and we are under steam, steering to the south-east, with a fine water sky ahead. Occasionally we are brought to a stop by a large floe, and have to seek our way round it; but, taking all things into consideration, our prospects are decidedly brighter than they have been for some time.

9 P. M.—I have just come below from witnessing one of the most glorious sunsets it has ever been my lot to see. For the last eight hours we have been boring through closely-packed ice, having been drawn into this extravagant proceeding in consequence of noticing a slight swell, which we, perhaps too sanguine, attribute to our proximity to the east water. No less than five bears were seen at different times this afternoon. I had a long chase on the ice after three of them, but could not get within eight hundred yards, and was not sorry to get on board the ship again, the ice being very thin and brashy, which, however, makes it easy to steam through.

Sunday, August 31st.—Last night, about 11 o'clock, to the great delight of every one, we emerged into the east water, and a course was shaped to take us down to Cape Farewell. Allowance, however, had not been made for indentations in the ice, and early this morning, a thick fog coming on, we found ourselves as badly off as ever, surrounded by loose pack. We took advantage of this delay to make fast to an iceberg, and took in about eight tons of ice, which will be equivalent to about six tons of water. Clearing up again at noon, we once more steamed into open water, and now really begin to look upon our troubles as being over and ourselves as actually homeward bound.

Monday, September 1st.—All doubts about being in the east water were entirely dissipated this morning. No pack-ice was to be seen anywhere, and we are now bowling along, at the rate of seven or eight knots, before a strong northerly wind and high sea. We passed several large icebergs during the day. This evening we are in about the latitude of Lievely, though many miles distant, being in the centre of Davis' Straits. If this wind lasts we shall be off Cape Farewell in three days. The boats have been denuded of their whaling gear, and are hoisted in-board, and we are now prepared in every way to cross the Atlantic.

Friday, September 5th.—Head winds, fair winds, and light winds, accompanied by a nasty jumble of a sea, have been our lot for the last three days, during which no sights. We imagine ourselves to be about one hundred miles north-west of Cape Farewell, which we hope to see to-morrow morning. I do not think we can be very far off land, as this evening I caught a small bird like a linnet, which had been blown off shore.

Tuesday, September 9th.—Of all the wretched, miserable, and comfortless days I have ever spent, the last four have eclipsed them. During the last twenty-four hours we have certainly had a fair wind, which in some degree compensates for the discomfort and cheerlessness of the previous days. "Battyfanging" about off Cape Farewell, with strong variable winds and a heavy cross sea, would not be cheerful on board any ship; however, it will all be at an end, I hope, in another week. The weather is getting oppressively warm, and the fire in the cabin has been allowed to go out for the first time since leaving Scotland. The thermometer has been as high as forty-seven degrees. We have caught two more small birds, one a linnet and the other a finch, which have been blown off to us from the Greenland shore during the late northerly winds. We are now, I am happy to say, well round the cape and in the broad Atlantic, and are hoping for a succession of westerly winds. It has been perfectly impossible to do anything during the last few days; even now the ship is so lively as to compel me to hold on to the inkstand with one hand while I write with the other.

Thursday, September 11th.—A fine clear day, with a light north-easterly wind, the ship progressing slowly at the rate of three knots an hour; but better that than being becalmed or having a head wind. Advantage was taken of the fine weather to wash and dry all the whale lines, which were towed overboard in a bight astern after the greasy places had been washed with fresh water and soft soap. On the arrival of the ship in harbour they will be removed to a loft on shore, where they will be thoroughly examined as to their fitness for another season. The harpoon guns are all carefully lubricated, sewn up in canvas, and stowed away for the winter. Men are going about the upper deck with paint brushes and paint of various colours, and giving a daub here and a patch there. This is what they call a priming coat, preparatory to painting. The funnel is being scraped at the same time, the dust from which, by sticking to the newly painted spots, gives them a somewhat novel appearance, resembling pepper and salt.

Saturday, September 15th.—Yesterday evening a fresh westerly wind sprang up, which has blown us at least two hundred miles nearer our destination during the last twenty-four hours. I much fear it will not continue, as it is inclined to head round to the southward and eastward, though at present we are going along at the rate of ten knots. Being fine, all hands were employed painting ship. The paint was laid on thickly and over dirt, without any regard to the blending of colour, of which there is great variety; red, blue, green, yellow, grey, black, and white, were put on wherever there was a vacant spot, the numerous shades, as the captain says, affording a pleasing relief. (?)

Sunday, September 14th.—We have lost our fine fair wind, which has been succeeded by an easterly one, dead ahead, leaving a heavy sea, in which we are tossing about in a most uncomfortable manner. Yesterday we were calculating upon getting into Dundee on Thursday, and now the chances are against our being there before this day week, unless the wind should shift or go down altogether, in which latter case I think we should have almost sufficient coal to steam the remainder of the distance.

Monday, September 15th.—This forenoon, having a light air from the southward, steam was raised, and we have been steaming easy since 11 A.M. One or two land birds were seen flying about the ship, though not sufficiently exhausted to alight. They were, however, a welcome sight, denoting the proximity of old England, distant between four and five hundred miles.

Wednesday, September 17th.—It is now a little past midnight, and we are just at the entrance of the Pentland Firth, bowling along before a strong westerly gale, under sail and steam, at the rate of ten or eleven knots an hour. It has been thick and raining all day, and for two days we have had no sights, notwithstanding which we have made a capital landfall, sighting the small islands of Barra and Rona at 3 P.M., after which the wind sprang up, gradually freshening to a gale, but a fair one. We sighted the light on Cape Wrath at half-past 7, and hope to get through the Firth by 3 A. M. If this wind lasts, there is a chance of getting to Dundee to-morrow night.

Thursday, September 18th.—Got through the Firth by 4 A. M.; a heavy following sea, and raining hard. At 11 we anchored off Peterhead, and landed our Shetland men. We weighed again at 7, and shall be in Dundee to-morrow morning.

Here ends the story of my whaling cruise in Baffin's Bay. Before, however, I take my leave, I must record my gratitude to those gallant seamen with whom it was my lot to be shipmates during my voyage in the good ship "Arctic." From the jovial and kind-hearted captain downwards, I always received the greatest respect and consideration. In fact, I was treated, if possible, with too great tenderness. When away in the boats assisting them in their calling, I was after a time not even permitted to pull an oar, but was always requested to take charge of the steer oar, a duty which they were aware occasioned me a great deal of pleasure.

Everything that was in any way conducive

to my wishes and amusement was immediately carried into effect; and though our living was a little of the roughest, still there was that air of genuine hospitality about it, which made our everlasting salt beef and biscuit a most cheering and appetizing banquet. My cruise in a Dundee whaler to Baffin's Bay will be one not quickly or easily forgotten by me, and I shall always look forward with pleasure to the chance (I am afraid a very poor one) of meeting my old shipmates again; nor will the survivors from the "Polaris," with whom it was my lot to associate for nearly three months, be forgotten. To Dr. Bessels I owe a great deal; he assisted me most materially in taking my observations, and in the collection of specimens which I brought to England. I trust he will reap the reward of his own exertions, whilst chief of the scientific department on board the "Polaris."

I have also to thank Dr. Soutter, the surgeon of the "Ravenscraig," for his valuable assistance in the illustration of this work.

If I have succeeded in interesting my readers in this attempt to depict life on board a whale-ship engaged in the fishery in Baffin's Bay, and the perils and dangers incidental to that occupation, and have also excited some little interest regarding those regions which McClintock, Osborn, and others of Arctic renown, have already so ably described, I shall feel myself amply rewarded.

" Mollies."

Appendix A.

Approximate Value and Size of Whales Captured during the Voyage of the "Arctic," in 1873.

Date.	Place.	Sex.	Length of Whalebone.		Weight of Whalebone.		Tons of oil.		Approximate value.	Remarks.
			Ft.	In.	Tons.	Cwt.	Tons.	Cwt.	£	
May 23	Off Resolution Island.	Male.	9	7	0	14	13	0	900	
June 15	Off Cape Byam Martin.	Female.	5	4	0	2½	3	10	220	
Do.	Do.	Male.	5	6	0	2½	3	10	220	
Do.	Do.	Female.	3	5	0	2	2	0	130	
Do.	Do.	Male.	7	0	0	4	5	10	330	
June 19	Off Cape Walter Bathurst.	Female.	11	7	1	5	19	0	1,450	
June 23	Do.	Female.	8	6	0	11	11	10	770	
Do.	Do.	Male.	7	0	0	5	6	0	380	
June 24	Off Cape Bowen.	Female.	10	4	0	19	11	0	950	
June 25	Do.	Male.	8	10	0	12	10	0	740	
June 28	Off Prince Regent Inlet.	Female.	9	2	0	13	12	0	850	
June 30	Do.	Female.	6	5	0	4	5	0	315	
Do.	Do.	Female.	10	4		0	20	0	1,360	With a young one or "sucker."
Do.	Do.	Male.	2	6	0	1		0	110	
Do.	Do.	Male.	4	0	0	2½	3	10	230	
July 5	In Barrow Straits.	Female.	11	9		5	20	0	1,500	With a "sucker."
Do.	Do.	Male.	6	6	0	5	7	0	420	
Do.	Do.	Female.		6	0	1½	4	0	200	
Do.	Do.	Male.	2	6	0	1½	3	0	160	
July 6	Off Point Sargent.	Female.	7	0	0	7	8	0	520	
July 9	Off middle ice, Baffin's Bay.	Male.	10	6	1	2	18	0	1,340	
July 10	Do.	Male.	11	0	1	0	13	0	1,060	
July 13	Do.	Male.	9	6	0	16	14	0	950	
July 20	In Admiralty Inlet.	Male.	5	3	0	2½	4	0	250	
July 29	In Prince Regent Inlet.	Female.	10	0	1	0	14	0	1,100	
Aug. 1	Do.	Male.	6	9	0	7	6	0	500	
Aug. 9	Off Cape Garry.	Male.	8	6	0	12	10	0	740	
Aug. 11	Do.	Female.	10	2		0	17	0	1,230	
Total	13 females and 15 males.		211	5	14	17	265	10	18,925	

Also 19 narwhals, 20 seals, and 12 bears, one of the latter being brought home alive, and which may now be seen at the Clifton Zoological Gardens.

NOTE ON THE BOILING DOWN OF THE BLUBBER.

THE blubber on board the whaling steamers arrives at Dundee in large tanks. It is there filled into casks, and taken to the boil-yard, to have the oil extracted. This operation is done by steam, in large coppers holding blubber sufficient to yield ten tons of oil. The seal blubber is so fresh when landed that it used to be kept stored in the boil-yard for six or eight weeks, until it was so decomposed that the oil might be easily taken out of it. But within the last year, the "Dundee Seal and Whale Fishing Company" have fitted up machinery for cutting and crushing the blubber, and can now boil it down as soon as it is landed. For some purposes the oil thus reduced is more valuable. After being boiled, the oil is allowed to settle in coolers, and is then run into large storing tanks, ready for delivery as required.

Appendix B.

ARCTIC PLANTS COLLECTED BY CAPTAIN A. H. MARKHAM, R.N., F.R.G.S.

1873.

Name.	Locality.
Ranunculus glacialis (*L.*)	Fury Beach. Elwyn Inlet.
Papaver alpinum (*L.*)	Fury Beach. Elwyn Inlet. Navy Board Inlet.
Lychnis apetala (*L.*)	Fury Beach.
Stellaria Edwardsii (*R. Br.*)	Elwyn Inlet. Fury Beach.
Dryas octopetala (*L.*)	Navy Board Inlet.
Saxifraga cæspitosa (*L.*)	Fury Beach.
„ nivalis (*L.*)	Fury Beach.
„ flagellaris (*Willd.*)	Fury Beach.
„ oppositifolia (*L.*)	Port Leopold. Elwyn Inlet.
Pedicularis hirsuta (*L.*)	Navy Board Inlet. Elwyn Inlet.
Juncus biglumis (*L.*)	Fury Beach.
Salix arctica (*R. Br.*)	14' S. of Cape Garry.
Alopecurus alpinus (*L.*)	Fury Beach.
Festuca ovina (*L.*) var.	6' S. of Cape Garry.
Pleuropogon Sabini (*R. Br.*)	Fury Beach.
Lichens.	
Platysma juniperinum (*L.*)	6' S. of Cape Garry. Fury Beach.
Alectoria ochroleuca (*Ehrh.*)	Fury Beach.

Plants from 82° N.; from the collection of Dr. Bessels.

Draba alpina (*L.*)
Cerastium alpinum (*L.*)
Taraxacum Dens-leonis (*Desf.* var.)
Poa flexuosa (*Wahl.*)

NOTE BY J. D. HOOKER, C. B., P. R. S.

CAPTAIN MARKHAM'S collection contains twenty species of flowering plants, including four collected by Dr. Bessels in the highest latitude from which flowering plants have hitherto been obtained, namely 82° N. They are *Draba alpina*, *Cerastium alpinum*, *Taraxacum Dens-leonis*, and *Poa flexuosa*. All of them are common Arctic plants, being found on both coasts of Greenland, as well as throughout the Parry Islands. Of the other species, collected by Captain Markham himself, the Arctic distribution is well known. None of them belong to the remarkable assemblage of Scandinavian plants which inhabit Greenland, and of which no member has hitherto been found on the eastern shores of Baffin's Bay. On the other hand, one of them is a member of that far smaller number which has never been found on the Greenland coast. This is the peculiar and beautiful little *Pleuropogon Sabini*, the only genus which is absolutely confined to the Arctic regions, and of which the solitary species is restricted in its distribution to the Arctic American Islands. It was discovered by Captain, now General Sir Edward, Sabine, in Melville Island, during Parry's first voyage in 1819-20, and is probably found in all the islands. Captain Markham's specimen was gathered on Fury Beach.

The other species call for no special remark. They are interesting as, in several cases, coming from places where the same plant had not previously been gathered. These localities are valuable, as completing our knowledge of the area inhabited by such species, though they do not materially enlarge it.

APPENDIX C.

LIST OF GEOLOGICAL SPECIMENS

COLLECTED BY

CAPTAIN A. H. MARKHAM, R.N., F.R.G.S.,

AND EXAMINED BY

R. ETHERIDGE, ESQ.,

MUSEUM OF PRACTICAL GEOLOGY.

UPERNAVIK.—1. *Syenite*, much resembling the Laurentian series of Cape Wrath (Sutherlandshire). 2. Crystals of felspar, also like those in the Sutherlandshire rocks. 3, 4. Quartz rocks.

ELWYN INLET.—Piece of quartz rock, and quartzite.

CAPE HAY.—Two pieces of limestone, extremely like that of the Durness in N. W. Sutherlandshire, of Llandeilo age (Lower Silurian).—Two specimens of *Saxicava rugosa*, from 150 feet above the sea level.

NAVY BOARD INLET.—Specimens of fundamental gneiss like that of Cape Wrath, hornblende rock, mica schist, quartzite, and magnesian limestone.

PORT LEOPOLD.—Syenite, felspar, and quartz, like the Cape Wrath rocks. An alternation of limestone and sandstone, probably Silurian. Gneissose rock, much the same as the fundamental gneiss of N. W. Sutherlandshire. A specimen showing annelide tracks, in fine-grained sandstone.

FURY BEACH.—Specimens of gneiss, hornblende, quartz, and gneissose rock, much like the fundamental series of Sutherlandshire. Argillaceous limestone, with the following fossils of the upper Silurian age:—*Favosites* (two specimens); *Athyris*, sp. (two specimens); *Holopella*, sp.

CAPE GARRY —Hornblende, and quartz rock stained red colour. Crystals of calcareous spar (carbonate of lime), concretionary limestone. Limestone containing several fossils of uncertain age. *Chonetes* and *Terebratula* of the Upper Silurian age.

Several of the specimens, having been picked up on the beach, are much waterworn.

Arctic Fox.

Appendix D.

LIST OF BIRDS SHOT.

Gyr Falcon (*Falco gyrfalco*)	1
Snow Bunting (*Emberiza nivalis*) and nest	1
Redcap, and Finches	4
King and Eider Duck (*Somateria spectabilis* and *mollissima*)	15
Long-tailed Duck (*Fuligula glacialis*)	10
Loom (*Uria Brunichii*)	800
Dovekey (*Uria grylle*)	100
Little Auks or *Rotches* (*Alca alle*)	500
Glaucous Gull (*Larus glaucus*)	2
Ivory Gull (*Larus eburneus*)	2
Fulmar Petrel or " Mollies " (*Procellaria glacialis*)	2
Arctic Tern (*Sterna arctica*)	2

Appendix E.

MEMORANDUM FOR THE ARCTIC COMMITTEE OF THE ROYAL SOCIETY.

June, 1873.

THE Arctic Committee of the Royal Geographical Society submitted to the committee appointed by the Council of the Royal Society to confer with them, the following memorandum on the subject of a renewal of Arctic exploration.

General Scientific Results.—The results of scientific importance to be derived from an examination of the immense unknown area round the North Pole are as numerous as the region to be explored is extensive. It may be shown that no such extent of unknown area, in any part of the world, ever failed to yield results of practical as well as of purely scientific value; and it may safely be urged that, as it is mathematically certain that the area exists, it is impossible that its examination can fail to add largely to the sum of human knowledge. Further, it is necessary to bear in mind that the Polar area is, in many most important respects, of an altogether special character, affording exclusive opportunities for observing the condition of the earth's surface, and the physical phenomena there to be seen, under certain extreme and singular circumstances, which are due to the relation of this area to the position of the axis of revolution of the terrestrial spheroid, and which have to be considered not only with reference to the present time, but to the earth's past history. It may be, therefore, received as certain that discoveries will be made in all branches of science, the exact nature of which cannot be anticipated. But there are also numerous objects, that have been stated and enumerated by the presidents and officers of the several scientific societies, the attainment of which make it desirable to despatch an Arctic expedition of discovery. These are as follows:—

Geography.—A geographical problem of great importance and interest will be solved by completing the circuit of Greenland, ascertaining the extent and nature of its northern coast, exploring the land to the westward, and discovering the conditions of land and sea in that portion of the unknown area.

Hydrography.—An Arctic expedition is a necessary complement to the expedition now investigating the ocean bottom in the middle and southern latitudes of the globe. The hydrography of the unknown seas has a most important bearing on the general question of oceanic currents, a question which is of practical consequence to navigation. Our knowledge of the general system of currents will be incomplete without an investigation of the currents, deep-sea temperatures, and soundings in the unknown area. Observations, at great depths, with the improved instruments now in use, would be of much value in connection with the like observations which are being carried on by the expedition now exploring the tropical seas.

Geodesy.—A series of pendulum observations at the highest latitude possible, following upon the series just completed in India, and made with the same instruments after verification at Kew, will be of essential service to the science of geodesy. Neither the data for forming a mathematical theory of the physical constitution of the earth, nor the means of testing such a theory, are complete without experimental determinations of the intensity as well as the direction of the force of gravity, and such observations would be especially valuable at the North Pole.

Meteorology.—Observations of the temperature of the sea at various depths; of temperature and pressure of the atmosphere; and of prevailing winds, with reference to currents, in very high latitudes, will form valuable contributions to meteorological science. The present state of meteorology requires a more thorough investigation of the motions of the earth's atmosphere than has yet been undertaken; and for this important object the less frequented parts of the earth's surface should be studied as well as the most frequented.

The climate of Europe in no small degree depends on the atmospheric conditions of the Polar area, in which the development of extremely low temperatures necessarily leads to corresponding extreme changes of pressure and other atmospheric disturbances, the effects of which are felt far into the temperate zone. For the satisfactory appreciation of these phenomena, a precise knowledge of the distribution of land and water within the Polar area is quite necessary, and any addition to our geographical knowledge of the Arctic region, accompanied by suitable observations of its meteorology, cannot fail to afford improved means of understanding the meteorology of our own country and of the earth generally.

Magnetism and Physics.—The extension of research into the phenomena of magnetism and atmospheric electricity, in the vicinity of the Poles, will necessarily be of much scientific importance; and generally, so far as the conditions of the climate and the means of an exploring expedition will permit, investigations in all branches of physics in the proximity of the Pole, where so many of the forces of nature operate in an extreme degree—either of excess or defect—will surely be followed by the acquisition of knowledge which can only be obtained in those exceptional localities.

The study of the aurora, which is among the most striking phenomena visible on our planet, is almost impossible in low latitudes; while the advance of spectrum analysis has given the means of determining the chemical elements involved, so that all that seems required here is the means of applying this description of observation; and this can only be got near the Pole.

The separation of the terrestrial lines from the truly solar ones in the solar spectrum, as seen from the earth's surface, is another important desideratum, inquiry into which can only be well pursued in high latitudes, where the path of the sun at low altitudes above the horizon gives opportunities for the necessary observations not to be secured elsewhere.

Geology.—A more complete investigation of the geology of the Arctic regions is extremely desirable, both for its scientific importance and the value of its practical results. The existence of Carboniferous, Jurassic, and Miocene rocks is known, but much is needed to be done to obtain complete collections of their organic remains. The existence of a true palæozoic coal formation has been determined, but we require to know its extent and composition.

One of the most interesting facts of late years acquired to geological science has been that of a luxuriant and highly organized vegetation, of the Miocene age, on the east coast of Greenland; a fact alluded to further on under the head of *Botany.* It is of great importance that some determinations based on fragments of leaves should be confirmed by the acquisition of more perfect foliage, as well as of seeds and fruits; such materials would be of great value in illustrating a flora which is in itself of much interest, but this interest is vastly increased when one realizes the important inquiries on which such knowledge would throw light. These inquiries are:—

1. The geographical distribution of the Miocene flora, as indicated by the agreements and differences between the Miocene plants of Arctic regions and of Central and Southern Europe.
2. The relation of the Miocene flora to previous and subsequent vegetations, and its bearings on the present geographical distribution of plants on the globe.
3. The evidence derived from these plants as to the physical conditions of the globe in past geological epochs.

It is certain that additional localities for fossil plants will be discovered, and of necessity additional species be brought to light, for, in the past, such remains have been found as far as explorers have penetrated.

From the important part extreme cold has of late years been found to have played in the last geological, or glacial, period, it would be of much value to have exact observations of the effects produced on the rocks by the intense cold of the northern regions; to ascertain the extent, height, and range of the glaciers; and to note their effects on the surface of the country and on the different classes of rocks. Again, it would be interesting to determine the extent of the river floods, and the depths of the channels they have excavated in the Arctic regions.

Another desirable object of the proposed Arctic expedition would be the investigation of the mollusca, not only of marine, but also of land and fresh water kinds. In a geological, as well as a zoological point of view, such an investigation would be especially valuable. The palæontological basis of the glacial epoch consists mainly in the identity of certain species which inhabit the Polar Seas, and are fossil in Great Britain and elsewhere. But such species may owe their present habitat and position to other than climatal causes, viz. to the action of marine currents. It is quite a mistake to assume that Arctic species are few in number; we know very little about them, because the exploration of the circum-polar seas by means of the dredge is so difficult. But the researches of the Scandinavian zoologists show that the Arctic marine invertebrate fauna is extremely varied and numerous. All fossils should be diligently collected, and their positions accurately noted. The condition and climate of the Arctic regions at the later geological periods may be thus ascertained, and a new chapter opened in the history of our globe.

The mineralogy of the Greenland continent is also important, and the discovery of new veins of cryolite and other valuable minerals is not improbable. Masses of meteoric iron have been recently discovered by the Swedish expedition, extending for a distance of not less than 200 miles; these require further study, and to have their position determined.

Botany.—The vegetation of the Arctic regions, in the opinion of Dr. Hooker, throws great light upon the geographical distribution of plants on the surface of the globe. On the return of Sir Edward Belcher's expedition from those regions, a series of rocks collected in the neighbourhood of Disco by his former fellow-voyager, Dr. Lyall, were placed in Dr. Hooker's hands, containing an accumulation of fossil leaves of plants totally different from any now growing in that latitude. These fossils he forwarded to Professor O. Heer, of Zürich, for investigation, who has brought forward the most convincing proofs that that latitude was once inhabited by extensive forests, presenting fifty or sixty different species of arborescent trees, most of them with deciduous leaves, some 3 in. or 4 in. in diameter—the elm, pine, oak, maple, plane, &c.; and what was more remarkable still, evidence of apparently evergreen trees, showing that these regions must have had perennial light. It seems extremely probable that the vegetation which belonged to the Miocene period extended over a large portion of the northern Arctic regions. It would be of great interest to ascertain whether such vegetation extends towards the Pole, and there is nothing that would give greater assistance in solving this problem than the proposed expedition along Smith Sound. Turning to the existing flora of Greenland, Dr. Hooker has pointed out that, though one of the most poverty-stricken on the globe, it is possessed of unusual interest. It consists of some 300 kinds of flowering plants (besides a very large number of mosses, algæ, lichens, &c.), and presents the following peculiarities:—1. The flowering plants are almost without exception natives of the Scandinavian peninsula. 2. There is in the Greenland flora scarcely any admixture of American types, which nevertheless are found on the opposite coast of Labrador and the Parry Islands. 3. A considerable proportion of the common Greenland plants are nowhere found in Labrador and the Parry Islands, nor, indeed, elsewhere in the New World. 4. The parts of Greenland south of the Arctic circle, though warmer than those north of it, and presenting a coast 400 miles in length, contain scarcely any plants not found to the north of that circle. 5. A considerable number of Scandinavian plants which are not natives of Greenland are nevertheless natives of Labrador and the Parry Islands. 6. Certain Greenland and Scandinavian plants, which are nowhere found in the Polar plains, Labrador, or Canada, re-appear at considerable elevations on the White, and the Alleghany, and other mountains of the United States. No other flora known to naturalists presents such a remarkable combination of peculiar features as this, and the only solution hitherto offered is not yet fully accepted. It is that the Scandinavian flora (which Dr. Hooker has shown evidence of being one of the oldest on the globe) did, during the warm period preceding the glacial—a period warmer than the present—extend in force over the Polar regions, including Greenland, the Polar American islands, and probably much now submerged land in places connecting or lying between Greenland and Scandinavia, at which time Greenland no doubt presented a much richer Scandinavian flora than it now does. On the accession of the glacial period, this flora would be driven slowly southward, down to the extremity of the Greenland peninsula in its longitude, and down to the latitude of the Alleghanies and White Mountains in their longitudes. The effect in Greenland would be to leave there only the more Arctic forms of vegetation, unchanged in habits or features, the rest being, as it were, driven into the sea. But the effect on the American continent would be to bring the Scandinavian flora into competition with an American flora that pre-occupied the lands into which it was driven. On the decline of the glacial epoch, Greenland, being a peninsula, could be re-peopled with plants only by the northward migration of the purely Scandinavian species that had been previously driven into its southern extremity; and the result would be a uniform Scandinavian flora throughout its length, and this an Arctic one, from north to south. But in America a very different state of things would supervene; the Scandinavian plants would not only migrate north, but ascend the Alleghanies, White Mountains, &c.; and the result would be that, on the one hand, many Scandinavian plants which had been driven out of Greenland, but were preserved in the United States, would reappear on the Parry Islands and Labrador, accompanied with sundry American mountain types; and, on the other, that a few Greenland-Scandinavian types which had been lost in the struggle with the American types during their northward migration, and which hence do not reappear in Labrador and the Parry Islands, might well be preserved in the Alleghanies and White Mountains. And, lastly, that a number of Scandinavian plants which had changed their form or habit during the migration in America in conflict with the American types, would appear in the Parry Islands as American varieties or representative species of Scandinavian plants.

Whether or no this be a true hypothesis, it embraces all the facts; and botanists look anxiously to further explorations in the northern parts of Greenland for more light on the subject, and especially for evidence of rising or sinking of the land in Smith Sound and the countries north and east of it, and for evidences of ancient connection between Greenland and Scandinavia; for observations on the temperature, direction, and depth of transporting currents in these seas, and on the habits of its ruminant migrating animals, that may have influenced the distribution of the vegetation by transporting the seeds. Such facts as those of the existence of ancient forests in what are now Arctic regions, and of the migration of existing floræ over lands now bound fast in perpetual ice, appear to some naturalists to call for vaster changes than can be brought about by a re-disposition of the geographical limits of land and sea, and to afford evidence of changes in the direction of the earth's axis to the plane of its orbit, and perhaps of variations in the ellipticity of the orbit itself.

It has thus been shown that much interest attaches to the Greenland flora, which is far from being exhausted. And besides these general questions, there are others respecting specific subjects, of which our existing knowledge is very imperfect. A great interest attaches to the minute forms of vegetable life which swarm in Polar areas, affording food to the cetaceæ and other marine animals, and which colour the surface of the ocean and its bottom likewise. Many of these forms are common to the Arctic and Antarctic seas, and have actually been far better studied in the latter than in the former sea. Of land plants the lichens and mosses require much further collection and study, and the Arctic marine flora is most imperfectly known. Ample collections of flowering plants should be made, with a view of testing the variability of species and their distribution; and observations on the means of transport of land plants by winds, currents, ice, and migrating animals, are very much wanted.

Zoology.—With regard to the specific results in zoology which may be expected from the proposed expedition, they are numerous and important. It is now known that the Arctic Ocean teems with life, and that of the more minute organized beings the multitude of kinds is prodigious: these play a most important part, not only in the economy of organic nature, but in the formation of sedimentary deposits, which in future geological periods will become incorporated with these rock-formations, whose structure has only lately been explained by the joint labours of zoologists and geologists.

The kinds of these animals, the relations they bear to one another, and to the larger animals (such as whales, seals, &c., towards whose food they so largely contribute), the conditions under which they live, the depths they inhabit, their changes of form, &c., at different seasons of the year and at different stages of their lives; and, lastly, their distribution according to geographical areas, warm and cold currents, &c., are all subjects of which very little is known.

With regard to the fish, mollusca, echinodermata, corals, sponges, &c., of the Arctic zones, those of Greenland alone have been explored with anything approaching to satisfactory results. A knowledge of their habits and habitats is much desiderated, as are good specimens for our museums. More important still would be anatomical and physiological investigations, and observations on those animals under their natural conditions.

With regard to the migrations of birds, Professor Newton, of

Cambridge, has drawn attention to some interesting points connected with the examination of the unknown area.

The shores of the British Islands, and of many other countries in the northern hemisphere, are annually, for a longer or shorter period, frequented by a countless multitude of birds, which, there is every reason to believe, resort in summer to very high northern latitudes, for purposes the most important; and, since they continue the practice year after year, they must find the migration conducive to their advantage. There must be some water which is not always frozen; secondly, there must be some land on which they may set their feet; and thirdly, there must be plenty of food, supplied either by the water or by the land, or by both, for their nourishment and that of their progeny.[1]

Ethnology.—The knowledge already acquired of the Arctic

[1] Professor Newton has furnished a short account of the movements of one species of birds—the knot—*Tringa canutus* of ornithologists. The knot is something halfway between a snipe and a plover. Examples of it are commonly to be seen in the cage at the southern end of the fish house in the Zoological Gardens, and may be seen there at the present time. Like many other kinds of birds belonging to the same group, the colour of its plumage varies most wonderfully according to the season of the year. In summer it is of a bright brick-red; in winter it is of a sober ashy-grey. Kept in confinement, it seldom assumes its most brilliant tints, but some approach to them is generally made. Now the knot comes to this country in vast flocks in spring, and, after remaining on our coasts for about a fortnight, can be traced proceeding gradually northwards till it takes its departure. People who have been in Iceland and Greenland have duly noted its appearance in those countries; but in neither of them is it known to tarry longer than with us—the summer it would there have to endure is not to its liking; and as we know that it takes no other direction, it must move further north. We then lose sight of it for some weeks. The older naturalists used to imagine it had been found breeding in all manner of countries, but the naturalists of the present day agree in believing that we know nothing of its nidification. Towards the end of summer it comes back to us in still larger flocks than before, and both old birds and young haunt our coasts till November; if the season be a very open one, some may stay later; but our winter, as a rule, is too much for it, and away it goes southwards, and very far southwards too, till the following spring. What has been said of the knot in the United Kingdom is equally true of it on the eastern shores of the United States. There it appears in the same abundance and at the

regions leads to the conclusion that the discovery of the unknown portion of the Greenland coast will yield very important results in the science of anthropology.

Light may not improbably be thrown upon the mysterious wanderings of those northern tribes, traces of which are found in every bay and on every cape in the cheerless Parry group, as well as up to the further point that has been reached beyond Smith Sound; and these wanderings may be found to be the most distant waves of storms raised in far-off centres, and among other races. Many circumstances connected with the still unknown northern tribes may tend to elucidate such inquiries.

There are other investigations which would undoubtedly yield valuable materials for the student of man. Such would be carefully prepared notes on the skulls, the features, the stature, the dimensions of limbs, the intellectual and moral state of individuals belonging to a hitherto isolated and unknown tribe; also on their religious ideas, on their superstitions, laws, language, songs, and traditions; on their weapons and methods of hunting; and on their skill in delineating the topography of the region within the range of their wanderings.

The condition of an isolated tribe, deprived of the use of wood or metals, and dependent entirely upon bone and stone for the construction of all implements and utensils, is also a

---

same seasons as with us, and its movements seem to be regulated by the same causes.

Hence we may fairly infer that the lands visited by the knot in the middle of summer are less sterile than Iceland or Greenland, or it would hardly pass over those countries, which are known to be the breeding-places of swarms of water-birds, to resort to regions worse off as regards supply of food. But the supply of food must depend chiefly on the climate. The inference necessarily is that, beyond the northern tracts already explored, there is a region which enjoys in summer a climate more genial than they possess. It would be easy to summon more instances from the same group of birds, tending to show that beyond a zone where a rigorous summer reigns there may be a region endued with a comparatively favourable climate. If so, surely the conditions which produce such a climate are worth investigating.

subject of study with reference to the condition of mankind in the stone age of the world; and a careful comparison of the former, as reported by explorers, with the latter, as deduced from the contents of tumuli and caves, will probably be of great importance in the advancement of the science of man.

Having thus epitomized the various scientific subjects which await investigation within the Polar area, it only remains to explain, from the knowledge and experience acquired up to the present time, why such researches can best be successfully accomplished by a naval expedition despatched under Government auspices, and secured as far as possible from failure or disaster by careful organization and good discipline.

It is now exactly a century since—in the year 1773—the British Government, moved by the Royal Society,[1] despatched

---

[1] The Royal Society took an active part in the furtherance of Arctic exploration up to the year 1845, and it is to be hoped that that eminent body will still persevere in a policy which has almost become traditional, and which has invariably been successful; for it cannot be said that any Arctic expedition despatched under their auspices ever returned empty-handed, or without an extension of our knowledge of the Polar seas, except that of 1845, when all the valuable results of three years' labour of Sir John Franklin's associates perished with that expedition.

In consequence of the representations contained in papers submitted by the Hon. Daines Barrington in 1773, the Royal Society resolved to apply to Lord Sandwich, then First Lord of the Admiralty, to obtain his Majesty's sanction for an expedition to be fitted out to explore the North Polar area. In a letter, dated January 19, 1773, the subject was recommended to Lord Sandwich, and it was urged that such discovery would be of service to the promotion of natural knowledge.

The wishes of the Council of the Royal Society were immediately complied with, and it was ordered that an expedition should be undertaken, "with every encouragement that could countenance such an enterprise, and every assistance that could contribute to its success." The command was given to Captain Phipps, afterwards Lord Mulgrave. The instructions were drawn up by Mr. N. Maskelyne, the Rev. H. Horsley, Mr. Cavendish, and Dr. Maty.

The comparative failure of Captain Phipps did not damp the ardour of the Royal Society. Early in 1774, the council minutes show that another expedition was frequently the subject of debate; and in February, 1774, a memorial was presented by the Royal Society to the Admiralty. This the first Polar expedition of modern times, under Captain Phipps, subsequently Lord Mulgrave, and in which expedition Lord Nelson served as a midshipman. But this, like all other expeditions sent *viâ* Spitzbergen, failed in its purpose of penetrating within the 80th parallel; and although Mackenzie and Hearn, on the American continent, just traced the two rivers which bear their names into the Arctic Sea, nothing in the last century was added to geographical knowledge within the Arctic zone to the rough outline of Baffin's Bay, as discovered by that great navigator in 1616; and, apart from that mere outline of Baffin's Bay and Spitzbergen, the entire area of the Arctic zone was a blank, so far as all human knowledge was concerned.

In the year 1818 the Royal Society, prompted by Sir Joseph Banks and Sir John Barrow, then Secretary to the Admiralty, took up actively the subject of Arctic exploration, and between

---

led to Captain Cook's attempt on the Pacific side; the expedition sailing in June, 1776.

In 1817 the council of the Royal Society resumed the consideration of the best means of prosecuting Arctic discovery, and a letter was addressed by Sir Joseph Banks to Lord Melville, dated November 20, 1817. A favourable reply was received on the 10th of December, in which it was announced that his Majesty's Government had deemed it their duty, in conformity with the suggestion of the Royal Society, to give orders for the fitting out of four suitable vessels, with a view to the important objects of Arctic discovery; two to proceed up Davis's Strait, and the other two along the east coast of Greenland to the northward. In a scientific point of view these expeditions were fruitful of results, including Sabine's magnetic observations.

On the return of Ross, another expedition was despatched in May, 1819, commanded by Parry; when Sabine again made valuable magnetic and pendulum observations.

In 1826 the council of the Royal Society again turned its attention to Arctic discovery, and Captain Parry proposed a plan to attempt to reach the North Pole by means of travelling with sledge-boats over the ice. Sir Humphry Davy, the President, wrote to Lord Melville, expressing the conviction of the council that Parry's expedition could not fail to afford several valuable scientific results, and to settle many important matters of scientific inquiry. Lord Melville replied "that, the council having no doubt balanced

that period and 1833, the successive expeditions of Franklin, Parry, Back, John and James Ross, Sabine, Buchan, Beechey, and Lyons added much to our geographical knowledge, and threw new light on the meteorology, botany, hydrography, terrestrial magnetism, zoology, and ethnology of a previously unknown portion of the earth's surface.

After the discovery of the exact position of the magnetic pole by Sir James Ross in 1831-33, Arctic exploration may be said to have paused; but it is worthy of remark that, during the fifteen years it had thus been actively pursued by seamen and travellers with the then imperfect means at command, no loss of life had occurred, although there had been occasionally more than two hundred men at a time employed upon these expeditions.

In 1845 the subject of Arctic research in various branches of natural science was again taken up by the Royal Society, and that year a fresh Arctic expedition was despatched by our Government, in which there were various persons eminent in science, under the command of Sir John Franklin.

It consisted of two sailing ships, with auxiliary steam-power

---

all the difficulties against the probable advantages, and having declared in favour of the expedition, I do not feel myself at liberty to withhold my assent to Captain Parry's earnest request." In a letter to the council, Captain Parry says that "the liberal and enlightened views of the council mainly led to the adoption of the enterprise by the Admiralty."

In 1839 the despatch of the Antarctic expedition commanded by Sir James Ross, though originally suggested by a committee appointed by the British Association, was urgently advocated by the president and council of the Royal Society, who threw themselves unreservedly and with their whole weight into the scale, with immediate and decisive effect. The council of the Royal Society then drew up a report containing a detailed account of every object of inquiry which should receive attention from the explorers.

In 1845 the council of the Royal Society again urged the importance of Arctic research, and their representations led to the despatch of the Franklin expedition, since which time no Government scientific expedition has been fitted out for the exploration of the unknown area round the North Pole.

of a very imperfect nature, and both in that respect, as well as in their general equipment, stores, and provisioning, they fell far short of what an Arctic expedition of the present day would have at command: but subsequent events reveal to us that this expedition succeeded in making one of the most remarkable Arctic voyages on record, and that they perished, after abandoning their ships, at a position near the entrance of the Great Fish River, where, had proper foresight been exercised, they could easily have been rescued. Subsequent experience has shown that the fatal omission which led to this catastrophe was the want of proper dépôts of provisions being arranged so as to cover the escape of the crews, in the event of disaster to the ships—a measure of precaution which, since that disaster, has always been carefully provided for in all subsequent expeditions with signal success. This expedition of Sir John Franklin in 1845 was the last *scientific expedition* sent by Great Britain into the Arctic regions. In 1848 the search for Franklin's expedition was pressed on the Government by the public, and from that date up to 1861—a period of thirteen years—was steadily persevered in, no less than fourteen public and private expeditions having gone and returned during that period. So far as the people in those expeditions were concerned, they all returned in safety, and the proportion of deaths from climate and disease was considerably less than the average death-rate of our naval seamen on any other service; and this in spite of the extraordinarily severe exposure and labour to which men and officers were subjected, by the novel introduction of sledge-travelling whilst the expeditions were frozen-in in winter quarters. Dr. Donnet, Deputy Inspector-General of fleets and hospitals, shows that at one period, out of 1,878 persons who wintered repeatedly in these expeditions, the death-rate was only 1·7 per cent., and states that the risk from climate and disease in a voyage to the Arctic seas "is not greater than that which a ship like the Challenger will incur in her voyage of discovery." These fourteen searching expeditions were equipped simply for the purpose of rescuing Franklin, and in no wise professed to be of an exploratory or scientific character; and it was only incidentally, and as a pure matter of individual zeal, that any one turned his attention to scientific observation, although, as a matter of fact, the various observations made by officers during their explorations contributed considerably not only to geographical, but to other branches of natural knowledge.

The general result pointed distinctly to the two following conclusions: that with the introduction of steam-power in Arctic ships, and the remarkable improvements in victualling them, navigation in the Polar seas had been rendered comparatively safe, and those maladies warded off from which seamen had suffered in ancient times. Further, that with proper organization and good discipline, double the work could be accomplished; whilst the men employed sought Arctic service as the most popular employment in the navy. The circumstance that for some years past the ordinary sailing whaler to Baffin's Bay has been entirely superseded by the fortified steam-ship, and that since this transition no fatal accident has occurred, but that these vessels annually reach a high northern latitude in pursuit of their calling and return with ease and safety, is one the significance of which cannot be over-stated.

On the solution of the fate of Franklin's expedition in 1861, Great Britain again withdrew from the field of Arctic research: but it was not so with other European nations. They, fired by the accounts of these different Arctic explorers, and of the honours reaped by British seamen and travellers, sought immediately to enter a field which had so redounded to our national honour; and Sweden, Germany, Austria, Russia, and notably America, year after year, made efforts to extend the area of human knowledge towards the North Pole, which, creditable and honourable as they were to those concerned, were undertaken with totally inadequate means and resources.

Under Dr. Kane and Dr. Hayes and Captain Hall, the Americans have attempted, with private expeditions, to emulate the achievements of the public ones of this country. The sufferings, the hardships, insubordination, and small results, in comparison with the expenditure and expectations, of these American private expeditions, fully confirm the opinions of all British Arctic authorities as to the necessity for the officers and seamen in such expeditions being always under naval control and discipline; and strengthen us in saying that no amount of private enterprise, enthusiasm, or funds will justify the risk to lives or the success of an expedition such as the Royal Geographical Society contemplates, except under Government auspices and Government control. That conceded, the safety of an expedition is comparatively guaranteed, so far as life is concerned, and its success for the objects set forth rendered doubly sure. It is contrary to fact, as has been alleged, that in public Arctic expeditions life has been sacrificed; and it is easy to show that the greater portion of the suffering and danger to which Arctic explorers have been subjected is owing to the want of organization and discipline incident to private expeditions, and to the expeditions being entrusted to unprofessional leaders. Moreover, it cannot be too strongly insisted upon that, with modern improvements and appliances, navigation in those seas has been made far more certain than it was in former years. That some risk may be incurred by individuals in prosecuting scientific research in an Arctic climate is not denied; but it may be confidently affirmed that no one who participates voluntarily in such an expedition would hesitate to incur such risks, and equally that life lost in the serious pursuit of knowledge is, to say the least, as worthily sacrificed as in other human occupations which involve similar dangers.

With these facts before us, we now turn to the subject of a fresh Polar expedition, of a purely scientific character, to deal with the points set forth in the first paragraph of this memorandum.

It will be seen, on reference to a circum-polar chart, that the entire area within the 80th degree of north latitude, except at two points—Parry's furthest in 1827, and the American explorations at Smith Sound—is an entire blank. In addition to this, there is a great area north of Behring's Straits, between long. 150° E. and 130° W., which is likewise unknown. The aggregate of these two areas around our northern Pole is not less than 2,400,000 square miles.

Since 1865 the council of the Royal Geographical Society have constantly had their attention turned to the desirability of extending their researches into this vast unknown region; and had they been justified in risking private expeditions upon such an enterprise, they might safely have appealed with success to their countrymen for funds and volunteers to undertake them; but they have, for reasons stated, preferred to urge such an undertaking on the Government, and in the same year a strong representation was made to the Duke of Somerset, then First Lord of the Admiralty, on the subject. At that time there was considerable divergence of opinion amongst English and other Arctic authorities as to the best route by which an expedition should be despatched for successful exploration within the unknown area around the North Pole, and Swedish and German expeditions were then making the attempt by way of Spitzbergen. His Grace declined to entertain the proposition until the results of those said expeditions were known.

In consequence of this view, the council of the Royal Geographical Society carefully watched the results of expeditions undertaken by foreign countries, in order to be in a position to recommend one route as undoubtedly the best, before again pressing the subject upon the attention of the Government. Eight years have now passed, and during that time additional experience has been accumulated by the Germans and Swedes, which has enabled the council to form an opinion that justifies the renewal of their representation made in 1865. The distinguished Arctic officers[1] who are members of the Geo-

[1] Sir George Back, Admiral Collinson, Admiral Ommanney, Admiral Richards, Sir Leopold McClintock, Admiral Sherard Osborn, Mr. Findlay, Mr. Clements Markham.

graphical Council, and who have carefully considered the evidence accumulated since 1865 in a special committee, are now unanimously of opinion that the route by Smith Sound is the one which should be adopted with a view to exploring the greatest extent of coast-line, and of thus securing the most valuable scientific results. They have recommended the Smith Sound route for the following reasons:—

1. That it gives a certainty of exploring a previously unknown area of considerable extent.

2. That it yields the best prospect of most valuable discoveries in various branches of science.

3. That, from the continuity of the land of Greenland and the Arctic archipelago southward from the 82nd parallel to the open sea in Baffin's Bay and Davis's Strait, it promises reasonable security for a safe retreat for the crews of an exploring expedition, should their ships be unable to be extricated from an advanced position, which, with steam-power, is a most remote possibility.

These opinions have been still further fortified by the recent report of the crew of the "Polaris," which ship, it appears, safely navigated up Smith Sound 250 miles beyond the point reached by Dr. Hayes's schooner in 1861, and traced the land on either hand as far as 82° 16′ N. She subsequently returned, and although a portion of her crew were separated from her, and took to an ice-field in 77° N., they drifted under the influence of the Polar stream down to a point in Labrador (where they were picked up this spring), a distance of 1,400 miles. This is the fifth occasion on which the Polar current through Smith Sound and Baffin's Bay has drifted vessels into the Atlantic; proving that the opening called Smith Sound is a channel with a constant current flowing southward from the unknown area.

The boat's crew from the "Polaris" reported open water at their furthest point to the north, in 82° 16′ N., a milder climate than has been found in more southern positions, and that terrestrial animal life abounded near their winter quarters, in 81° 38′ N., including musk oxen—a point the importance of which cannot be overrated.

The Admiralty have the means, by referring to past records and living authorities, of laying down clearly and economically all the requirements for such an expedition as is contemplated. It is therefore unnecessary in this memorandum to enter into any lengthened detail on the subject. But we may say that in general terms we only seek that it should consist of two moderate-sized screw-steamers, one to be stationed at some distance within the entrance of Smith Sound, the other to advance, as far as possible, to the northward (preserving communication with the dépôt vessel), from which point sledge parties would start in the early spring, and explore the unknown region in various directions, whilst the scientific staff on board the respective ships, being in near proximity to the land, would be able to prosecute researches both on shore and by means of the ice on the sea. The advanced parties would be in such a position as to be able to fall back upon the consort, at her station near the entrance of Smith Sound. Thence, in the improbable event of accidents, the whole expedition could retreat to the Danish settlements in Greenland, as has been before done.

In conclusion, we may be allowed to add that the council of the Royal Geographical Society has never appealed to the Government to undertake enterprises which are of a nature to admit of being carried out by private enterprise. In almost every part of the unknown regions of the globe their emissaries have been and are abroad, and at the present time they have on hand *two* expensive and difficult explorations in the interior of Africa. But, for Arctic exploration, the conditions under which the investigations must be made, for reasons already explained, are such that they can best be conducted through the instrumentality of a Government expedition.

CHISWICK PRESS:—PRINTED BY WHITTINGHAM AND WILKINS, TOOKS COURT, CHANCERY LANE.

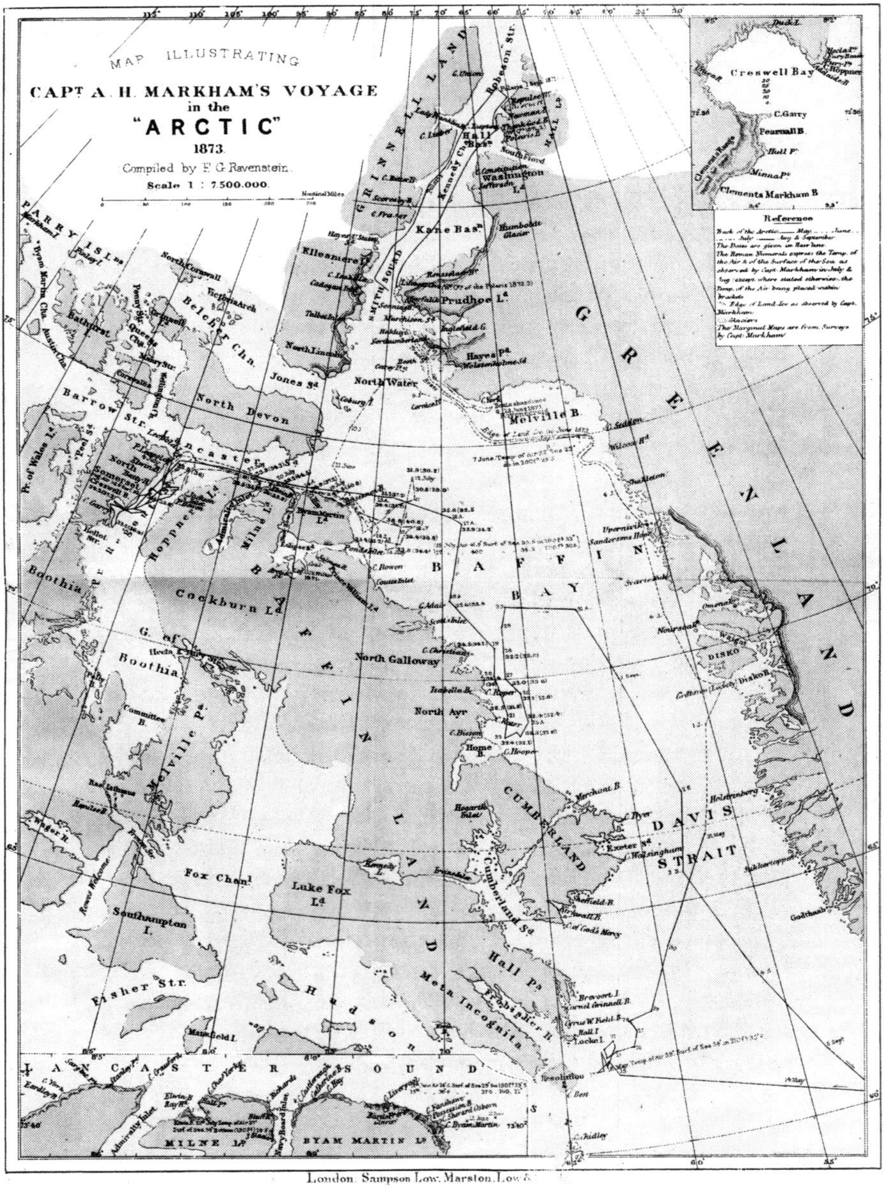

The material originally positioned here is too large for reproduction in this reissue. A PDF can be downloaded from the web address given on page iv of this book, by clicking on 'Resources Available'.

CPSIA information can be obtained at www.ICGtesting.com
Printed in the USA
LVOW06s1142290414

383590LV00001B/150/P

9 781108 072007